In this series original contributions of physical research in all fields are contained, such as

— research monographs
— collections of papers to a single topic
— reports on congresses of exceptional interest for physical research.

This series is aimed at promoting quick information and communication between physicists of the various special branches.

In diese Reihe werden Originalbeiträge zu allen Gebieten der physikalischen Forschung aufgenommen wie

— Forschungsmonographien
— Sammlungen von Arbeiten zu einem speziellen Thema
— Berichte von Tagungen, die für die physikalische Forschung besonders aktuell sind.

Die Reihe soll die schnelle Information und gute Kommunikation zwischen den Physikern der verschiedenen Fachgebiete fördern.

Manuscripts
in English comprising at least 100 pages and not more than 500 pages can be admitted to this series. With respect to a quick publication the manuscripts are reproduced photomechanically. Authors who are interested in this series please turn directly to the Akademie-Verlag Berlin. Here you will get more detailed information about the form of the manuscripts and the modalities of publication.

Manuskripte
in englischer Sprache, die mindestens 100 Seiten und nicht mehr als 500 Seiten umfassen, können in diese Reihe aufgenommen werden. Im Interesse einer schnellen Publikation werden die Manuskripte auf fotomechanischem Wege reproduziert. Autoren, die an der Veröffentlichung entsprechender Arbeiten in dieser Reihe interessiert sind, wenden sich bitte direkt an den Akademie-Verlag Berlin. Sie erhalten dort genauere Informationen über die Gestaltung der Manuskripte und die Modalitäten der Veröffentlichung.

I. M. Tsidilkovski

Gapless Semiconductors

Physical Research

Volume 7

Gapless Semiconductors — a New Class of Materials

by
I. M. Tsidilkovski

Gapless Semiconductors — a New Class of Materials

by
I. M. Tsidilkovski

With 36 Figures

Akademie-Verlag Berlin
1988

Author:

Isaak Michailovich Tsidilkovski
Institute of Physics of Metals
Academy of Sciences of the USSR, Sverdlovsk

И. М. Цидильковский
Бесщелевые полупроводники — новый класс веществ
Издательство Наука, Москва 1986

Translated from the Russian by
Anatoli Prokopyevich Savarnitsyn

Reproduction of the original author's manuscript

ISBN 3-05-500450-7

Erschienen im Akademie-Verlag Berlin, DDR-1086 Berlin, Leipziger Str. 3—4

Lizenznummer: 202 · 100/436/88
Printed in the German Democratic Republic
Gesamtherstellung: VEB Druckerei „Thomas Müntzer", 5820 Bad Langensalza
LSV 1185
Bestellnummer: 763 755 1 (9079)

02700

PREFACE

On reading the title of the book, the reader may query whether the gapless (or zero-gap) semiconductors are really a new class of materials. It would seem that much has already been written about these semiconductors. Even in newspapers. But twenty years ago the concept of a "gapless semiconductor" did not yet exist. This term came into scientific use in 1964-1965, although the gapless state was first discovered in 1955. To answer the notion of "a new class", the materials that belong to it should meet apparently four (or at least three of the four) criteria: (1) the features peculiar to these materials radically distinguish the latter from those known (in our case, from the "normal" semiconductors and metals); (2) a short span of time has elapsed since this group of materials was discovered; (3) the degree to which the gapless semiconductors have been studied does not correspond to what theses like to qualify as "comprehensively" or even "completely"; (4) the wide scientific community is inadequately conversant with these materials, as few reviews, if any, are available and there are no monographs at all. Since two decades in an age of a scientific and technological revolution is a fairly long period, the gapless semiconductors may, strictly speaking, claim to conform to the first, the third, and, to some extent, the fourth criterion (in 1983 a monograph by Dornhaus, Nimtz, and Schlicht was published which was devoted to narrow-gap semiconductors and, at the same time, gave some treatment to zero-gap semiconductors as well). For a wide range of laymen, the present text will be the first book on the physics of gapless semiconductors, while hundreds of monographs, textbooks, and popular books deal with the physics of "normal" semiconductors.

If one compares the paces at which semiconductor physics and zero-gap semiconductor physics advanced within an equal space of time, say twenty years (from 1948 to 1968 and from 1965 to 1985 respectively), he should acknowledge that the physics of zero-gap semiconductors advanced at an appreciably slower pace. In this respect, too, the materials under discussion are still "new", i.e., satisfy the second criterion as well. The more rapid development of the physics of "normal" semiconductors is, in our view, due to the fact that the semiconductors, which in 1947 gave the world a transistor, have led actually to the contemporary scientific and technological revolution. The "merits" of the gapless semiconductors, on the basis of which infrared radiation detectors and other controled electronic devices have been made, are undoubtedly more modest.

At the same time the reader should be aware that "a slow development" is a very relative notion. Many hundreds of papers have been published on gapless semiconductors in the past two decades, and the author was faced with the challenge of selecting material for a small-sized book.

What do we look for in a book that we, intrigued by its title, take for the first time into our hands? When scientific literature is concerned, we naturally look for new information. If that is a textbook, it is essential that experimental facts and theoretical constructions be treated clearly and coherently, for a textbook should teach. Of a monograph one requires that the exposition be profound and the references and bibliographic information be amply supplied. The book you are holding in your hands may evidently not serve as a textbook; nor is it a monograph in the strict sense of the word.

For lack of space it has not been possible to include some, in the author's view, very important problems in which considerable progress has been made in recent years. The present book may give a layman an idea about the subject, the most significant results, unresolved problems, and prospects.

The text consists of two large chapters, of which one treats the band structure of zero-gap semiconductors with a strictly periodic crystal lattice and the other is devoted to the effect of impurities on the band structure and discuss some transport characteristics. The book deals only with materials whose gapless state arises from lattice symmetry (fundamental band degeneracy). Typical representatives of this class of materials are gray tin and mercury telluride. No consideration is given to materials whose gapless state is due to a particular value of the crystalline potential acting on the electron (accidental band degeneracy). Among these materials are, for example, bismuth-antimony alloys , ternary solid solutions of lead, tin, and tellurium. The two large chapters are prefaced with a small one which qualitatively describes the fundamentals of the band theory of solids that are requisite for further reading.

Any province of physics can be expounded rigorously only by invoking to a smaller or greater extent the mathematical machinery. Here we content ourselves with mainly final formulas for analyzing particular physical characteristics and phenomena. Of course, whereever possible, the reader is told what considerations underlie the derivation of a formula or how it is obtained. Sometimes the author had to resort to the locution "It may be shown ...", leaving off screen the answer to the question "in what fashion?".

From the viewpoint of practical applications, the most important property of the gapless semiconductors is the extremely high sensitivity of their electronic system, chiefly that of the electron and hole concentrations, to external effects such as temperature, pressure, dc electric and magnetic fields, and light. It is for this reason that a variety of electronic devices have been created on the basis of gapless semiconductors. The problem of the practical utilization of zero-gap semiconductors is also given due attention.

There are some sections in the book which require a larger amount of sophistication on the part of the reader than the other. Here belong, for example, problems about the role of the electron-electron interaction and of the electronic phase transitions. A more penetrating insight into the problems treated in the book may be gained from the reviews and monographs cited below. In these the reader will find all of the most important original publications.

Whom is the book expected to appeal to? To be able to read it, one should be acquainted with the essentials of quantum mechanics (wavefunction, Schrödinger equation), statistical physics (distribution function), and with the fundamentals of the electronic theory of semiconductors, especially with those of band theory. A knowledge at a level of the courses taught at the corresponding departments of technical colleges and, all the more, universities is quite sufficient. When writing this book, the author assumed that it would appeal to physicists unconcerned with research of gapless semiconductors, physical chemists and chemists, electronic engineers, industrial engineers who make and study semiconductor materials, and certainly university and college students majoring in solid state physics. Also, this book may possibly be of interest to specialists in the physics of gapless semiconductors.

It is a pleasant duty to acknowledge here the assistance of Vyacheslav Reprintsev and Natalya Stripkina in preparing the English manuscript.

I.M. Tsidilkovski

REFERENCES

1. B.L. Gelmont, V.I. Ivanov-Omskii, and I.M. Tsidilkovski, (1976) Uspekhi Fiz. Nauk 120, 337.

2. N.N. Berchenko and M.V. Pashkovskii, (1976), Uspekhi Fiz. Nauk 119, 223.

3. I.M. Tsidilkovski, (1982), Band Structure of Semiconductors, Pergamon Press, Oxford.

4. I.M. Tsidilkovski, G.I. Harus, and N.G. Shelushinina, 1985, Impurity States and Electron Transport in Gapless Semiconductors, Adv. in Phys. 34, No 1, 43-174.

5. I.I. Lyapilin and I.M. Tsidilkovski, 1985, Uzkoschelevye polumagnitnye poluprovodniki, Uspekhi Fiz. Nauk 146, vyp.1, 35-72.

6. R. Dornhaus, G. Nimtz, and B. Schlicht, 1983, Narrow-Gap Semiconductors, Springer Tracts in Modern Physics, No 98, Heidelberg, New York.

7. N.B. Brandt and V.V. Moshalkov, 1984, Semimagnetic Semiconductors, Adv. in Phys. 3, 193-255.

CONTENTS

CHAPTER 1

ELEMENTS OF THE BAND THEORY OF SEMICONDUCTORS

CHAPTER 2

BAND STRUCTURE OF GAPLESS SEMICONDUCTORS

CHAPTER 3

IMPURITIES IN GAPLESS SEMICONDUCTORS

CONCLUSION

CHAPTER 1

ELEMENTS OF THE BAND THEORY OF SEMICONDUCTORS

1.1. The Problem of the Electronic States in Solids

A crystalline solid is an assembly of two subsystems: an atomic (ionic) one, which forms the skeleton of the crystal lattice, and an electronic one. The major features peculiar to semiconductors, primarily the high sensitivity of these materials to external effects which has determined their role in modern technology, are due to the electronic subsystem. When a crystal is in the equilibrium state, the interatomic spacings are such that the atomic-shell electron wavefunctions overlap, the atomic energy levels of electrons broaden to form energy bands, and the electrons become capable of moving over the free states in those bands. The wavefunctions of the outermost atomic shells overlap so strongly that the corresponding energy bands turn out to be sufficiently wide (≈1 to 10 eV) and the electron is able to move through the entire crystal. These outer-shell electrons make up a system of conduction electrons whose wavefunctions are non-zero in the bulk of the crystal. The electron wavefunctions of deeper atomic shells overlap not as strongly, the corresponding bands are narrower, and the electrons are virtually localized at their nuclei. In most of the cases, especially in studying charge transfer processes, a solid may therefore be viewed as a collection of ions (nucleus plus inner shell electrons) and conduction electrons.

The states of conduction electrons and ions are determined by the Schrödinger equation

$$\mathcal{H}\Psi = E\Psi \tag{1.1}$$

with $\mathcal{H}$ being the Hamiltonian (Hamilton's operator), and Ψ and E the wave eigenfunctions and energy eigenvalues of the electronic and ionic system. The Hamiltonian $\mathcal{H}$ contains terms that specify the kinetic of ions and electrons, and also involves terms that characterize the potential energies of the Coulomb electrostatic interaction of all charged particles among themselves. Solving the Equation (1.1) would, in principle, permit one establish the main (electrical, magnetic, thermal, and other) properties of a solid. However, rigorous solution of the Schrödinger equation, which describes the interaction of a giant number of particles ($\simeq 10^{23}$), is practically impossible. It is therefore the object of theory to provide sufficiently well-founded and, at the same time, sufficiently simple approximate methods of solving the Equation (1.1), techniques that would enable one to calculate quantities observed in experiment and to account for the behavior of the observables when subject to temperature, pressure, electric and magnetic field, etc.

The first simplification of the problem of finding the energy spectrum of electrons in a crystal is based on the large difference that exists between ion mass M and electron mass m_0. It is therefore assumed that the velocity of travel of ions is much less than the electron velocity. The band theory of solids considers the motion of electrons in a field of immobile ions and neglects processes that are associated with the alteration of the equilibrium configuration of ions.

Even now, however, the Hamiltonian contains terms that describe the electron-electron interaction, i.e., the problem continues to be a many-electron problem. A further approximation consists in that the potential interelectron interaction energy in the Equation (1.1), $\frac{1}{2}\sum_{i\neq j} \frac{e^2}{|\vec{r}_i - \vec{r}_j|}$ ($\vec{r}_i$ and $\vec{r}_j$ are the coordinates of electrons), is replaced by some effective potential $\sum_i U_{eff}(\vec{r}_i)$ which specifies the averaged action of all electrons on a given electron. This permits the Equation (1.1) for the system of ions and electrons to be reduced to a one-electron equation of the form

$$\left(- \frac{h^2}{2m_o} \nabla^2 + U(\vec{r})\right) \psi(\vec{r}) = \mathcal{E}\, \psi(\vec{r}) \tag{1.2}$$

with m_o being the free electron mass, and $U(\vec{r})$ the potential energy of an electron with the coordinates $\vec{r} = (x,y,z)$ in-the averaged field of all the other electrons and all the ions in the crystal. The quantity $U(\vec{r})$ is a periodic function of the coordinates and is called the crystalline potential, the kinetic energy operator ∇^2 being equal to $\frac{\partial^2}{\partial x^2} + \frac{\partial^2}{\partial y^2} + \frac{\partial^2}{\partial z^2}$.

Finding the eigenfunctions ψ and energy eigenvalues $\mathcal{E}$ in the Equation (1.2) reduces to two problems: (1) determining the periodic potential $U(\vec{r})$ and (2) solving the Equation (1.1) for a given potential. So far there is unfortunately no rigorous method of constructing the crystalline potential and one has to content himself with approximations. The most important and frequently used simplification is to replace the nonlocal exchange interaction operator (which depends on the coordinates $\vec{r}_1$ and $\vec{r}_2$ of the interacting electrons) by some averaged local operator that depends on only one coordinate, say, $\vec{r}_1$. Another simplification is that the Coulombian part of the crystalline potential is assumed to be equal to the sum of the Coulombian parts of the atomic potentials. This is equivalent to the assumption that the electron density of the crystal is equal to the sum of the electron densities of the atoms of the crystal.

The second problem, i.e., the solution of the Schrödinger equation at a given potential, has been explored much more fully. A number of methods have been developed which allow the electronic spectra of semiconductors and metals to be calculated to an accuracy approaching that attained by experimental means (insofar as energies go, the accuracy ranges between 0.01 and 0.1 eV). A detailed discussion of all the computational methods available lies outside the scope of this book. Here we wish to treat briefly, and only qualitatively, two approaches to the problem.

1.2. The Tight-Binding Approximation

One of the approaches to the problem of electronic states in solids, called the tight-binding approximation, consists in what follows. First and foremost, we are interested in what will happen to the electronic levels of isolated atoms when the latter, as they come closer together, form a crystal. Let the atoms be initially arranged as a space lattice of required composition and symmetry and be spaced macroscopic distances apart. If we uniformly compress such a lattice while preserving geometric similitude, we will obtain an ideal (defect-free) crystal. As the atoms come closer together the interatomic interaction enhances; it is this interaction that is responsible for the transformation of the electron energy spectrum of an isolated atom to the electron spectrum of the crystal. A fundamental assumption in the tight-binding approximation is that the electronic state of an isolated atom alters insignificantly when a crystal forms, and the wavefunction of an electron in a crystal may be represented as linear combination of electron wavefunctions for a certain atomic state, which is specified by a set of quantum numbers: principal quantum number n, orbital quantum number l , and magnetic quantum number m .

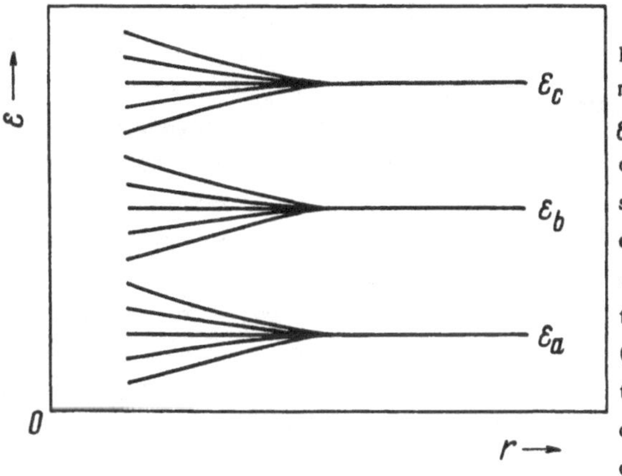

Fig.1. Diagram of crystal energy band formation from atomic levels of electrons when atoms come closer together

According to the Pauli exclusion principle, not more than two electrons may be present on each atomic level ε_a, ε_b, ε_c, etc. (Fig.1). In a regular crystal lattice composed of N atoms spaced a large distance apart, each electronic state (wavefunction) φ_a, φ_b, φ_c, etc., which is nondegenerate in the atom, becomes N-fold degenerate (the degeneracy of a state means that to different wavefunctions there corresponds the same energy). This so-called permutational degeneracy, which does not arise from interaction, implies that an electron of energy ε_i may be in any one of the N lattice atoms located far from each other. As the atoms come closer together the wavefunctions of the electrons start overlapping to form electron wavefunctions of the crystal. The permutational degeneracy is lifted and each electron wavefunction in the crystal has an energy level of its own.[1)] Put another way, the previously N-fold degenerate levels split up and form energy bands that consist of N closely adjacent levels. Between the bands there are zones of forbidden energies, which are said to be energy gaps or forbidden bands. Inasmuch as each atomic level 2s,3p,4d, etc.[2)] in the crystal has a band of its own, one may talk of 2s-, 3p-, 4d-

1) Some states, degenerate in the atom, may remain degenerate in the crystal as well. In a free atom many states are degenerate owing to a high degree of free-space symmetry. Thus in the central symmetry force field, 2l+1 states (with neglect of spin) correspond to the energy of an electron with the principal quantum number n, which differ in orbital angular momentum orientation. This (2l+1)-fold degeneracy arises from the fact that the electron energy in a central-symmetry field is independent of orbital angular momentum orientation. As the crystal field symmetry is lower than the field symmetry of a separate atom, many degenerate states of the atom should split up. Part of the states may remain degenerate since the crystalline field itself possesses a definite symmetry.

2) The ciphers 2,3,4,... denote the principal quantum number n, and the letters s,p, d,... refer to values of the orbital quantum number 0,1,2,... . Wavefunctions of the s-type (l=0) possess spherical symmetry, i.e., do not depend on the polar φ and azimuthal α angle of the spherical system of coordinates. Three p-type functions exist for l=1; these are p_x, p_y, and p_z, which have the shape of dumb-bells and correspond to the same energy (triply degenerate state) and depend in different ways on the angles φ and α .

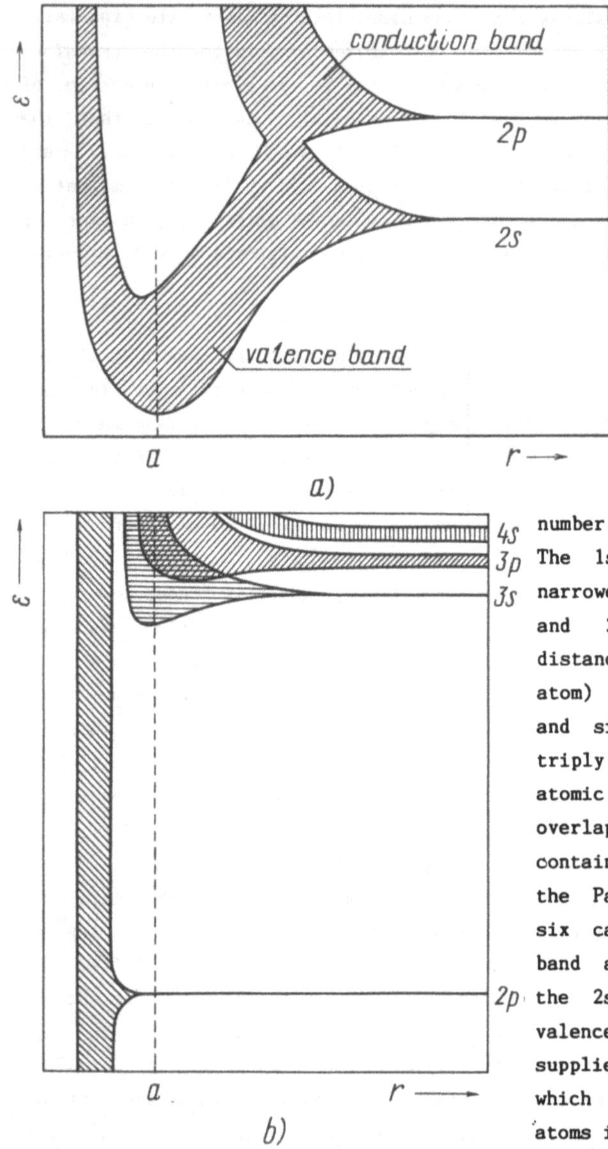

Fig.2 Energy bands of diamond (a) and sodium (b). a is the equilibrium interatomic distance (lattice spacing) in the crystal.

bands, etc. This band classification proves justified only in the case when the bands, made up of atomic levels of different types (s,p,d,etc.), do not overlap.

Figure 2a) presents the 2s and 2p energy bands of carbon. The latter has the electronic configuration $1s^2 2s^2 2p^2$, where the numerals above indicate the number of electrons in each of the state. The 1s levels form a band that is narrower than and lies lower than the 2s- and 2p-bands. When the interatomic distances are large, two electrons (per atom) may lodge in the low-lying 2s-band and six electrons can reside in the upper triply degenerate 2p-band. As the interatomic spacing is decreased the two bands overlap and each of them turns out to contain four levels per atom. According to the Pauli exclusion principle, two of the six carbon atom electrons fill the 1s band and the other four electrons fill the 2s band. This band is called the valence band since the four electrons supplied by each atom are precisely those which form covalent bonds between carbon atoms in the diamond lattice.[3)]

At temperature T = 0 K the upper band of diamond, portrayed in Figure 2a), is free from electrons; it is called the conduction band. If a valence band electron emerges in

3) Since the 2s-shell is wholly filled, it would seem that carbon should be divalent. However, all the data of organic chemistry bear testimony to the fact that carbon is tetravalent. To form four valence bonds four unpaired electrons are needed. Such a configuration may be realized if one of the 2s-electrons passes to the 2p-shell, i.e., if the carbon atom passes to the excited $1s^2 2s\, 2p^3$-state. Thus a tetravalent sp^3-state should arise: One valence electron is in the s-state and three valence electrons are in the p-state (p_x,p_y,p_z). In reality it turns out that valence bonds are produced not by s- or p-type wavefunctions but by some hybrid electron states which are linear combinations of one 2s- and three 2p-states. Such hybrid wavefunctions assure a larger overlap than pure s- and p-functions, providing a gain in energy. Spatially, the four hybrid equivalent sp^3-bonds are directed from the center of a regular tetrahedron to its vertices. Put another way, carbon forms tetrahedral bonds. Tetrahedral structure is inherent in the methane molecule CH_4, diamond, silicon, germanium, and gray-tin crystals.

this band , say, when the temperature is increased or on exposure to a light quantum of sufficient energy to overcome the forbidden band, that electron will be able to move through the crystal even when subject to the weakest electric field. As the electron is accelerated by the electric field, the energy of the electron increases and the electron should pass on to higher-lying energy levels. In the conductions band, the adjacent energy states[4)] are free and the electron, when subject to an electric field, can occupy these.

The situation is different in the valence band, where all the states at 0 K are occupied. The electron is incapable of acceleration by an electric field, since all the levels that lie higher are occupied. It is only very strong fields, capable of imparting to the electron an energy larger than the energy gap between the bands, or, as is customary to say, inducing an interband breakdown, that can "make" valence band electrons participate in the electrical conduction of the crystal.

The arrangement of the bands (Fig.2b)) for sodium with the atomic spacing a corresponding to a stable equilibrium of the crystal is quite different from that for diamond. The bands originating from the 3s, 3p, and higher-lying atomic levels overlap. Of the eleven sodium atom electrons, ten atomic electrons fill the 1s-, 2s-, and 2p-bands; the eleventh electron emerges in the 3s-band, which turns out to be half-filled. This situation, with half of the valence band being populated with electrons, is typical of all alkali metals. In alkaline-earth elements the number of electrons is just sufficient for the valence band to be filled completely, as in diamond. But the valence band overlaps with an empty band, so alkaline-earth elements are metals. The level that delimits the occupied and free states is called the Fermi level. In sodium, just as in all the metals, the weakest electric field induces a current, since there is no energy gap between the occupied and free states, that is to say, the Fermi level lies in a band of allowed energies.

1.3. The Nearly Free Electron Approximation

In the tight-binding approximation one is concerned with changes in the state of electrons bound to individual atoms, i.e., changes that result from the interaction between atoms. An alternative approach to the problem of electronic states in a solid is what is known as the nearly free electron approximation. This approach proceeds from an assembly of free electrons and considers changes in the behavior of these electrons when placed into the crystal lattice. The crystalline field, described by the potential $U(\vec{r})$ in the Equation (1.2) is viewed as a small perturbation of the state of a free electron. Put another way, the potential energy of an electron, $U(\vec{r})$, in the periodic field of the crystal is small compared to its kinetic energy $\mathcal{E} = \hbar^2k^2/2m_0$, where $\vec{k}$ is the electron wavevector ($\hbar\vec{k}$ is the momentum). The function $\mathcal{E}(\vec{k})$ for a free electron is a parabola.

4) The valence band width W is normally equal to several electron volts. It hence follows that the distance between the levels in the band is vanishingly small: $W/N \simeq 10^{-23}$ eV, i.e., the band is practically quasicontinuous.

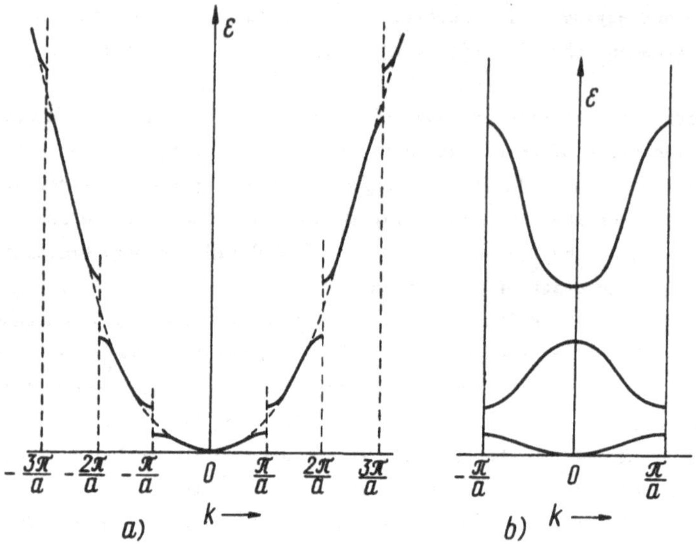

Fig.3. $\mathcal{E}(k)$ relation in the nearly free electron approximation. One-dimensional case (a): $-\infty < k < +\infty$; mapping of the $\mathcal{E}(k)$ function on the first Brillouin zone (b): $-\pi/a \le k \le +\pi/a$

For a free electron (U=0), the solution of the Schrödinger Equation (1.2) has the form of plane waves

$$\psi = ce^{i\vec{k}\vec{r}}. \tag{1.3}$$

In the nearly free electron method the wavefunction of an electron in the crystal is represented as a series of plane waves (Fourier series). We will come to this representation of the wavefunction if we expand the potential $U(\vec{r})$ in a Fourier series. Substitution of these expressions for the wavefunctions and potential into the Schrödinger equation permits one, on the assumption that U is small, to find the energy as a function of the wavevector. It turns out that the $\mathcal{E}(\vec{k})$ relation is a modified parabolic function, which is typical of free electrons (one-dimensional case) (see Fig.3). For the values $k = \pm q\pi/a$, with a being the lattice period and q = 1,2,3,..., the curve has discontinuities; when far from these $\mathcal{E}(k)$ values, the k curve is a parabola. The magnitude of the discontinuities depends on the potential U. Thus the regions of allowed energies, i.e., the bands, are separated by forbidden areas - gaps. Both approximations have led us to the electron spectrum band structure, which is a consequence of crystal lattice periodicity.

The interval of variation of k between $-\pi/a$ and $+\pi/a$ (correspond to the low-lying allowed energy band) is called the first Brillouin zone, the interval from $-2\pi/a$ to

$-\pi/a$ and from $+\pi/a$ to $+2\pi/a$ the second Brillouin zone,etc. In the case of two dimensions the Brillouin zone have the shape of polygons, in the three-dimensional case they are polyhedra. For a simple cubic lattice with a period (lattice spacing) a, the first Brillouin zone is a cube with an edge $2\pi/a$, the volume of which is equal to $8\pi^3/\Omega_o$, with $\Omega_o = a^3$ the volume of a unit cell of the crystal. The volume of the first Brillouin zone is equal to $8\pi^3/\Omega_o$ for any lattice, where $\Omega_o = \vec{a}_1[\vec{a}_2, \vec{a}_3]$ ($\vec{a}_1$ stands for three lattice spacings). By analogy with the usual three-dimensional vector space $\vec{r}=(x,y,z)$, one often introduces the reciprocal wavevector space $\vec{k} = (k_x, k_y, k_z)$. This permits one to introduce the concept of a reciprocal lattice with periods $2\pi\vec{b}_1$, $2\pi\vec{b}_2$, and $2\pi\vec{b}_3$, the vectors $\vec{b}_i$ being related to the direct-lattice periods by the equations

$$\vec{b}_1 = \frac{[\vec{a}_2, \vec{a}_3]}{\Omega_o}, \quad \vec{b}_2 = \frac{[\vec{a}_3, \vec{a}_1]}{\Omega_o}, \quad \vec{b}_3 = \frac{[\vec{a}_1, \vec{a}_2]}{\Omega_o}.$$

For the reciprocal lattice, just as for the direct lattice, one introduces three mutually orthogonal symmetry axes: [100], [010], and [001].

To this point we were considering the variation of k between 0 and $\pm\infty$. However, the periodicity of the crystal allows the energy to be expressed in terms of the reduced wave vectors that lie outside the first Brillouin zone: $-\pi/a \le k \le +\pi/a$ (Fig.3b)). This is a direct consequence of the periodicity of the wavefunction ψ: When k is replaced by $k \pm 2\pi q/a$, the function ψ retains its form. The energy of state may now be described by a wavevector varying within the confines of the first Brillouin zone and by the quantum number n that specifies the energy band to which the state under consideration pertains.

Thus for a given wavevector $\vec{k}$ the Schrödinger Equation (1.2) has an infinite set of solutions, i.e., wavefunctions $\psi_{n\vec{k}}(\vec{r})$ and energy $\mathcal{E}_n(\vec{k})$ values. Any electron state in the crystal is completely characterized by the reduced wavevector $\vec{k}$ and energy band number n.

We wish to enlarge upon some properties of the wavefunctions $\psi_{n\vec{k}}(\vec{r})$ and energies $\mathcal{E}_n(\vec{k})$, which arise from crystal periodicity. This lattice periodicity determines primarily the translational symmetry of the potential energy of an electron, $U(\vec{r})$, i.e., the invariance of the potential energy under translation of the electron from cell to cell: $U(\vec{r}+\vec{R}) = U(\vec{r})$, where $\vec{R} = R_1\vec{a}_1 + R_2\vec{a}_2 + R_3\vec{a}_3$ is the lattice vector; when displaced by this vector, the lattice goes into itself (R_i are integers). It is manifest from the Schrödinger Equation (1.2) that the Hamiltonian $\mathcal{H}$ in this case possesses translational symmetry, too. This signifies that in the lattice the points $\vec{r}$ and $\vec{R}+\vec{r}$ are physically equivalent. The wavefunctions at these points may therefore be expected to coincide to the constant factor $\psi(\vec{r}+\vec{R}) = c\,\psi(\vec{r})$, with $|c_{\vec{R}}|^2 = 1$. It may be shown that the $c_{\vec{R}}$ dependence of $\vec{R}$ should have the form $c_{\vec{R}} = e^{i\vec{k}\vec{R}}$. We have arrived at an important consequence of translational lattice symmetry: $\psi(\vec{r}+\vec{R}) = e^{i\vec{k}\vec{R}}\psi(\vec{r})$. For an electron with the wavevector $\vec{k}$ in the n-th band of the crystal, the wavefunction meeting this requirement (which is also satisfied by the free-electron wavefunction $\psi_{\vec{k}}(\vec{r}) = \exp(i\vec{k}\vec{r})$) may be represented as

$$\Psi_{n\vec{k}}(\vec{r}) = u_{n\vec{k}}(\vec{r})\, e^{i\vec{k}\vec{r}} \tag{1.4}$$

where the function $u_{n\vec{k}}(\vec{r})$ possesses lattice periodicity:

$$u_{n\vec{k}}(\vec{r} + \vec{R}) = u_{n\vec{k}}(\vec{r}) \quad .$$

The function $u_{n\vec{k}}(\vec{r})$ satisfies the equation by substituting the Bloch function (1.4) into the Equation (1.2)

$$\left(\frac{\hat{\vec{p}}}{2m_o} + \frac{h}{m_o} \vec{k}\,\hat{\vec{p}} + U(\vec{r})\right) u_{n\vec{k}}(\vec{r}) = \mathcal{E}_n(\vec{k})\, u_{n\vec{k}}(\vec{r}) \quad , \tag{1.5}$$

with $\hat{\vec{p}} = i\hbar\nabla$ the electronic angular momentum operator $\vec{p} = \hbar\vec{k}$.

If we write the Schrödinger Equation (1.2) for the complex conjugate function $\psi^*_{n\vec{k}}(\vec{r})$, then, with the potential $U(\vec{r})$ being real-valued, we come to the same Equation (1.5), in which the vector $\vec{k}$ is replaced by $-\vec{k}$. It hence follows that

$$\mathcal{E}_n(-\vec{k}) = \mathcal{E}_n(\vec{k}) \quad . \tag{1.6}$$

The Equation (1.6) tells us that the energy of an electron in the crystal is an even function of the wavevector.

By contrast with the wavevector of a free electron, the wavevector $\vec{k}$ of an electron in the crystal has been defined ambiguously. It follows from lattice periodicity that $\psi_{\vec{k}+\vec{g}}(\vec{r}) = (\vec{r})$, where $\vec{g} = 2\pi(g_1\vec{b}_1 + g_2\vec{b}_2 + g_3\vec{b}_3)$ is the reciprocal-lattice vector (g_i are integers; when displaced by the vector $\vec{g}$, the reciprocal lattice goes into itself). A consequence of the periodicity of the wavefunction in the reciprocal lattice is that all the physically observable quantities that depend on $\vec{k}$, in particular the energy $\mathcal{E}_n(\vec{k}+\vec{g}) = \mathcal{E}_n(\vec{k})$, are periodic. In view of its ambiguity , the quantity $\vec{p} = \hbar\vec{k}$ is said to be the crystal momentum (quasimomentum). To unambiguously define the wavevector that corresponds to a given wavefunction, it suffices to restrict our attention to the first Brillouin zone ($-\pi \leq \vec{k}\vec{a}_i \leq \pi$, $i = 1,2,3$) since the reciprocal-lattice points, which differ by the vector $\vec{g}$, are physically equivalent (for each point $\vec{k}'$ outside the first Brillouin zone there is an equivalent point $\vec{k}$ within the zone).

Let us see now what shape constant-energy ($\mathcal{E}$ = const) surfaces have in $\vec{k}$ space. In the one-dimensional case (we choose the x axis) two points $k_x = \pm\sqrt{2m_o\mathcal{E}_o}/h$ correspond to a given energy value $\mathcal{E}_o$, as is clear from Figure 3. In the two-dimensional case, electrons of energy $\mathcal{E}_o$ possess k values that correspond to a circle in the k_xk_y plane with the center at $\vec{k} = 0$: $2m_o\mathcal{E}_o/\hbar^2 = k_x^2 + k_y^2$. Such circles are easy to obtain by rotating the parabolic segments of the $\mathcal{E}(k)$ curves, depicted in Figure 3, about the $\mathcal{E}$ axis. What corresponds to the energy $\mathcal{E}_o$ in the three-dimensional case is the surface in k space which are spheres: $2m_o\mathcal{E}_o/\hbar^2 = k_x^2 + k_y^2 + k_z^2$. In fact, constant-energy surfaces may be ellipsoidal or have complicated shapes. As a matter of fact, it was tacitly assunmed in our reasoning that the mass of an electron in the crystal is equal to the mass of a free electron, m_o. Since the perturbation of electronic states which is caused by the potential U is normally large, it often proves impossible, in contrast to the free-electron case, to describe the $\mathcal{E}(\vec{k})$ relation for an electron in the crystal by a quadratic function. In those cases where this function is quadratic, the coefficient of proportionality between $\mathcal{E}$ and k^2 is other than $1/2m_o$. In semiconductors we are usually concerned with energy band edges (extrema), that is, the conduction band minimum and valence band maximum, as the charge carrier occupancy of the bands is inconsiderable, viz., many orders of magnitude lower than that in metals.

1.4. Effective Mass and the Density of States

Let a wavevector $\vec{k}_o$ correspond to the band extremum. The expansion of the energy in a power series of $\vec{k}$ up to k^2 order terms in the vicinity of the extreme points $\vec{k}_o$ has the form

$$\mathcal{E}_n(\vec{k}) = \mathcal{E}_n(\vec{k}_o) + \frac{1}{2}\sum_{\alpha,\beta} \left(\frac{\partial^2 \mathcal{E}_n(\vec{k})}{\partial k_\alpha \partial k_\beta}\right)_{\vec{k}=\vec{k}_o} (k_\alpha - k_{o\alpha})(k_\beta - k_{o\beta}) \quad , \tag{1.7}$$

where α and β run independently over the values 1,2,3(x,y,z). The first-order derivatives $(\partial \mathcal{E}_n / \partial k_\alpha)_{k_\alpha = k_{o\alpha}}$ near the extreme point are equal to zero.

The second-order derivatives involved in the Equation (1.7) generate a symmetric tensor of rank two. The $\mathcal{E}_n(\vec{k})$ relation, normally referred to as the dispersion relation (1.7), may be written in a form close to the dispersion relation $\mathcal{E}(\vec{k}) = \hbar^2 k^2/2m_0$ for a free electron:

$$\mathcal{E}_n(\vec{k}) = \mathcal{E}_n(\vec{k}_o) + \frac{\hbar^2}{2}\sum_{\alpha,\beta}^{3} \left(\frac{1}{m_{\alpha\beta}}\right)_n (k_\alpha - k_{o\alpha})(k_\beta - k_{o\beta}) \quad , \tag{1.8}$$

where the quantities

$$\left(\frac{1}{m_{\alpha\beta}}\right)_n = \frac{1}{\hbar^2} \left(\frac{\partial^2 \mathcal{E}_n(\vec{k})}{\partial k_\alpha \, \partial k_\beta}\right)_{\vec{k}_o} \tag{1.9}$$

form an inverse effective mass tensor. The components of this tensor have the dimensionality of inverse mass.

Solving the Equation (1.5) for the function $u_{n\vec{k}}(\vec{r})$ in the vicinity of the extreme point $\vec{k}_o$ up to quadratic terms in $(\vec{k}-\vec{k}_o)$ enables us to find the dispersion relation $\mathcal{E}_n(k)$, which may be written in the form of the Equation (1.8), where

$$\left(\frac{1}{m_{\alpha\beta}}\right)_n = \frac{\delta_{\alpha\beta}}{m_o} + \frac{2}{m_o^2} \sum_{n' \neq n} \frac{p^\alpha_{nn'}(\vec{k}_o)\, p^\beta_{n'n}(\vec{k}_o)}{\mathcal{E}_n(\vec{k}_o) - \mathcal{E}_{n'}(\vec{k}_o)} \tag{1.10}$$

In this equation $p^\alpha_{nn'}(\vec{k}_o)$ stands for the α-th matrix element component of the momentum operator $\hat{\vec{p}}$ at $\vec{k}=\vec{k}_0 : \vec{p}_{nn'}(\vec{k}_0) = \int u^*_{n\vec{k}_0} \hat{\vec{p}} u_{n'\vec{k}_0} \, d\vec{r}$; $\delta_{\alpha\beta} = 1$ at $\alpha=\beta$ and $\delta_{\alpha\beta}=0$ at $\alpha \neq \beta$. The denominator of the second term in the Equation (1.10) involves the energy gaps between the bands at point $\vec{k}_o$. If only two bands are important (the other bands are widely separated in energy), the Equation (1.10) for the electron in the first band may be rearranged in the form

$$\left(\frac{1}{m_{\alpha\beta}}\right)_1 = \frac{\delta_{\alpha\beta}}{m_o} + \frac{2}{m_o^2} \frac{p^\alpha_{12}(\vec{k}_o)\; p^\beta_{21}(\vec{k}_o)}{\mathcal{E}_g(\vec{k}_o)} \tag{1.11}$$

where the gap $\mathcal{E}_g(\vec{k}_o)=\mathcal{E}_1(\vec{k})-\mathcal{E}_2(\vec{k}_o)$ is direct. The gap is said to be direct if the band edges (the lowest conduction band minimum and the highest valence band maximum) are located at the same Brillouin zone point $\vec{k}_o$. Henceforth both direct and indirect gaps will be labeled by the same symbol $\mathcal{E}_g$.

If the tensor (1.9) is reduced to the basis axes, the dispersion relation (1.8) becomes

$$\mathcal{E}_n(\vec{k}) = \mathcal{E}_n(\vec{k}_o) + \frac{h}{2}\sum_{\alpha=1}^{3}\left(\frac{1}{m_\alpha}\right)_n (k_\alpha - k_{o\alpha})^2 \tag{1.12}$$

The second term on the right-hand side of the Equation (1.12) has the form of the kinetic energy of an electron that possesses different masses in different directions. The quantities m_α are called effective masses. The constant-energy surfaces in $\vec{k}$ space $\mathcal{E}_n(\vec{k})$=const have, according to the Equation (1.8), the shape of ellipsoids. The parity of energy with respect to the wavevector (see Eq.(1.6)) means that the isoenergetic surfaces $\mathcal{E}_n(\vec{k})$=const are symmetric with respect to the point k=0, i.e., always possess a center of symmetry. The extrema of the bands, i.e., of the functions $\mathcal{E}_n(\vec{k})$, always lie on the symmetry axes <100> ([100], [010], [001]), <110>, <111> or at the highest-symmetry point $\vec{k}$=0.

If the principal values of the tensor (1.9), $(m_\alpha^{-1})_n = \hbar^{-2}(\partial^2\mathcal{E}_n/\partial k_\alpha^2)_{\vec{k}_o}$ are equal in magnitude, we are in a position to introduce the scalar effective mass $m=m_1=m_2=m_3$. Setting $\vec{k}_o$ = 0 and taking $\mathcal{E}_n(0)$ to be zero energy, we come to the simplest form of the quadratic isotropic dispersion relation:

$$\mathcal{E}_n(\vec{k}) = \frac{\hbar^2 k^2}{2m_n} . \tag{1.13}$$

The dispersion relation (1.13) differs from the relation for a free electron in only that it involves the effective mass rather than the mass m^*. Isoenergetic surfaces are spheres.

We wish now to explore the relation between the electron concentration n and the wavevector $\vec{k}_F$ which corresponds to the Fermi energy $\mathcal{E}_F=\hbar^2k_F^2/2m$, i.e., the boundary energy that separates occupied states from free ones. In the crystal of dimensions L_1, L_2, and L_3, an electron performs a finite motion, i.e., its coordinates are bounded by the values of L_1, L_2, and L_3. In keeping with the Heisenberg indeterminacy principle, the wavevector components of the electron can be determined to an accuracy that does not exceed $\Delta k_i=2\pi/L_i$ (i=1,2,3). This implies that only determinate points in k space correspond to allowed electronic states.

We refer each of these points to a cubic cell $\Delta k_1\Delta k_2\Delta k_3 =(2\pi)^3/L_1L_2L_3=8\pi^3/\Omega$, with Ω being the volume of the crystal. Thus we have broken the k space up into cells, the number of which is equal to the number of allowed states. For a macroscopic crystal volume Ω these cells are extremely small. The number of cells, i.e., the number of states in a volume element $dk_1dk_2dk_3$ of $\vec{k}$ space per unit volume of the crystal is equal to $\frac{2dk_1dk_2dk_3}{8\pi^3} / \Omega \cdot \frac{1}{\Omega} = \frac{2dk_1dk_2dk_3}{8\pi^3}$. The factor 2 present in the numerator indicates that to each $\vec{k}$ correspond states with opposite spin orientations. If the electrons fill in $\vec{k}$ space a sphere of radius k_F (i.e., of volume $4\pi k_F^3/3$), the electron concentration n is equal to $(4/3)\pi k_F^3 2/8\pi^3$. Hence

$$k_F = (3\pi^2 n)^{1/3} \quad , \quad \text{or} \quad \mathcal{E}_F = \frac{\hbar^2}{2m}(3\pi^2 n)^{2/3} . \tag{1.14}$$

Our intension now is to find the expression for the density of states, $\rho(\mathcal{E})$. The number of cells that get into the spherical layer between the spheres of radii k and k+dk is (with allowance made for spin) equal to $k^2\Omega dk/\pi^2$. In the wavevector variation interval between k and k+dk there are $k^2 dk/\pi^2$ cells per unit volume of the solid. The number of states over the energy range from $\mathcal{E}$ to $\mathcal{E}+d\mathcal{E}$ in unit volume is $\rho(\mathcal{E})d\mathcal{E}=\rho(k)dk=k^2dk/\pi^2$, where $\rho(\mathcal{E})$ is the density of states corresponding to the energy $\mathcal{E}$. On writing $dk/d\mathcal{E}$ for the quadratic isotropic dispersion relation $\mathcal{E}=\hbar^2k^2/2m$, we find

$$\rho(\mathcal{E}) = \frac{2^{1/2}\ m^{3/2}}{\pi^2\ \hbar^3}\ \mathcal{E}^{1/2} . \tag{1.15}$$

In the case of ellipsoidal constant-energy surfaces the expression for $\rho(\mathcal{E})$ involves, instead of m, the density of states effective mass $m_d=M^{2/3}(m_1m_2m_3)^{1/2}$, where M is the number of ellipsoids available for the first Brillouin zone, and m_1, m_2, and m_3 are the effective masses which correspond to the three principal axes of an ellipsoid.

Our goal now is to consider a nondegenerate energy band that is composed of atomic s functions. Let the edge of the band, which we will denote by the index 1, be located at $\vec{k}=0$. The matrix elements of the momentum operator $\vec{p}_{12}(0)$ are nonzero if the index 2 pertains to the edge of a triply degenerate p-type band: $p_{12}^x(0)=p_{12}^y(0)=p_{12}^z(0)\equiv p_{12}(0)$. Such a situation occures in cubic crystals. It will be assumed that all the other bands are located far from the two bands of interest, 1 and 2. The constant-energy surfaces of an s-type band are spherical. Suchlike surfaces are specified by the scalar effective mass m_1, which, according to the Equation (1.11), is defined by the expression

$$\frac{1}{m_1} = \frac{1}{m_o} + \frac{2}{m_o^2}\frac{|p_{12}(0)|^2}{\mathcal{E}_g} . \tag{1.16}$$

The matrix elements $\vec{p}_{nn'}$ enter into the expression for the light absorption coefficient which is due to interband transitions. This allows one to determine the quantities $\vec{p}_{nn'}$ from optical measurements and to calculate the effective masses. The formula (1.16) is often employed in a somewhat different form, with p_{12} being replaced by the matrix element P which is related to p_{12} by the equation $|P|^2 = \hbar^2/m_o^2|p_{12}|^2$. Instead of P, one sometimes introduces the energy $\mathcal{E}_p=2m_o|P|^2\hbar^2$. The Equation (1.16) then assumes the form

$$\frac{1}{m_1} = \frac{1}{m_o}\left(1+\frac{\mathcal{E}_p}{\mathcal{E}_g}\right) . \tag{1.17}$$

At the lower energy band edge, where $\mathcal{E}_n(\vec{k})$ reaches a minimum, the second-order derivatives $(\partial^2\mathcal{E}_n(\vec{k})/\partial k^2_\alpha)_{\vec{k}_o} > 0$ and the effective masses m_α are positive. At the upper band edge, where $\mathcal{E}_n(\vec{k})$ has a maximum, $m_\alpha < 0$. The effective mass is one of the most important characteristics of the electron in the crystal. On the one band, the effective mass determines the dispersion relation near the band edge; on the other, it charaterizes the dynamic properties of the electron in the crystal. When the electron is subject to a force $\vec{F}$ of an electromagnetic (but not inertial) nature, the change in the state of the electron in the crystal, i.e., its motion, is described by an expression similar in form to the second Newton's law for a free particle

$$\frac{d(\hbar \vec{k})}{dt} = F . \qquad (1.18)$$

For the isotropic quadratic dispersion relation, specified by the scalar effective mass m which is determined by relation $m\vec{v}=\hbar\vec{k}$ (with $\vec{v}$ being the velocity of an electron), we obtain from the Equation (1.18)

$$m \frac{d \vec{v}}{dt} = \vec{F} . \qquad (1.19)$$

In the case of an anisotropic quadratic dispersion relation (ellipsoidal isoenergetic surface), the force component along the α axis is

$$\frac{d v_\alpha}{dt} = \sum_\beta \frac{1}{m_{\alpha\beta}} F_\beta , \qquad (1.20)$$

where F_β is the force component along the β direction (α,β = 1,2,3).

If the dispersion relation is nonquadratic, the effective mass (1.10) may not be introduced rigorously and the equation of motion may not be reduced to the form of the Equation (1.20). However, for a nonquadratic isotropic dispersion relation, with the isoenergetic surfaces being spherical, the effective mass concept, although depending on energy $\mathcal{E}$ (or the modulus $|\vec{k}| = k$), proves often to be instrumental in interpreting phenomena associated with the change of electronic states. In this case the effective mass m, involved in the expression for kinetic, optical, and other effects, is defined by the first-order energy derivative with respect to the wavevector

$$\frac{1}{m(k)} = \frac{1}{\hbar^2 k} \frac{\partial \mathcal{E}(k)}{\partial k} . \qquad (1.21)$$

The negative sign of the effective mass at the upper band edge means that the electron is accelerated in a direction opposite to the direction of the acting force $\vec{F}$. This fact, surprising at first sight, results from simple allowance for the combined action of a periodic lattice field and an externally applied force on an electron that possesses wave properties. The electron motion near the upper band edge, where $m<0$, i.e., the motion of an assembly of electrons of a lightly filled band, may be described as a motion with respect to a relatively small number of unoccupied states in that band, that is, holes. The latter are viewed as quasiparticles with a positive charge +e and positive effective mass m_h:

$$\left(\frac{1}{m_h}\right)_{\alpha\beta} = -\left(\frac{1}{m_e}\right)_{\alpha\beta} = -\frac{1}{\hbar^2}\frac{\partial^2 \mathcal{E}(\vec{k})}{\partial k_\alpha \partial k_\beta} . \qquad (1.22)$$

What is implied by quasiparticles is carriers of dynamic properties in a system of interacting particles of a solid. Quasiparticles are generated by the cooperative motion of coupled particles rather than the motion of a separate particle, i.e., a structural unit of a substance (an ion, an electron). If there are several types of interaction, several types of quasiparticles may exist as well. In a solid these are conduction electrons (for brevity referred to as electrons), holes, phonons (thermal vibration quanta), excitons (electrons and holes coupled by the Coulomb electrostatic interaction), etc. In respect of dynamic properties, a hole, just as an electron at the lower band edge, is a "normal" quasiparticle, i.e., it is accelerated in the direction of the acting force.

In an intrinsic semiconductor a light quantum whose energy exceeds the gap $\mathcal{E}_g$ between the conduction band and the valence band generates simultaneously an electron in the conduction band and a hole in the valence band; the electron and hole thus produced are capable of transferring the charge through the crystal (intrinsic conductivity). Evidently, equal amounts of electrons n and holes p arise in this case: $n=p=n_i$. If the gap between the bands is larger than the mean thermal energy of a charge carrier ($\mathcal{E}_g \gg k_BT$, with k_B being the Boltzmann constant), the intrinsic charge carrier concentration is

$$n_i \sim \exp(-\mathcal{E}_g/2k_BT) .$$

1.5. Impurity Levels. Statistics

In what fashion do impurity atoms and other lattice defects affect electron and hole concentration? Stranger atoms may either replace lattice atoms (substitutional impurities) or wedge themselves in between them (interstitial atoms, or interstitials). The most important effect that impurity atoms bring about is the formation of new energy levels in the electron spectrum of the crystal. Let, for example, a germanium lattice atom be replaced by a group V atom, say, an antimony atom. Four of the five valence electrons of the Sb atoms and four neighbouring Ge atoms participate in the formation of covalent bonds similar to those existing in a diamond-like Ge lattice. The fifth electron that does not take part in bond formation (as there are no vacant sites in the neighborhood) is attracted by relatively weak Coulomb forces on Sb. The energy of this bonding will be evaluated somewhat later in the text. If a Ge atom is replaced by a group III atom, for example In, three In valence electrons form bonds with three neighbors in the lattice; the fourth bond is lacking in an electron - a hole arises (in the language of chemical bonds the shortage of an electron in a bond is equivalent to the formation of a hole in the valence band in terms of band theory). Due to Coulomb attractive forces, the hole is retained by the In atom. If the hole is remote from the

In atom (i.e., the electron is attracted to In), the fourth covalent bond of In will be filled and the In atom will turn into a negatively charged ion. For this reason, the group III elements in Ge, Si, etc. are called acceptors, for they readily attract a valence band electron to form electron vacancies in the valence band. The group V elements, which readily furnish an electron to the conduction band, are said to be donors. The conductivity due to conduction band electrons or valence band holes excited off donors or acceptors, respectively, is referred to as the extrinsic (impurity) conductivity.

Where should the extra electronic levels due to donors and acceptors be located? In the case of Sb in Ge the fifth electron should be in a higher energy state than the electrons that produce valence bonds, i.e., belong to the valence band. On the other hand, since this electron is bound to the Sb atom, the energy of the electron of interest should be much less than that of the lower edge of the "free" states, i.e., the conduction band energy. Put another way, the additional electronic donor level due to the Sb atom should be located in the forbidden energy band. The same applies to the acceptor level (see Fig.10b)).

The ionization (binding) energy of the fifth Sb electron, i.e., the energy needed to remove it from the Sb atom and to transfer it into the conduction band, may be estimated if we note that this problem has some analogy with the problem of hydrogen atom ionization. A point of dissmilarity is that in the first case the electron revolves in an orbit around the Sb atom and one has to allow for the effect of the lattice somehow or other. Since the electron is bound to the Sb atom loosely enough, the orbit of the electron encompasses many atoms of the lattice, which may therefore be viewed to a good approximation as a continuous medium with a dielectric constant κ . The lattice effect in such a case reduces primarily to the electron-Sb-atom binding energy decreasing by a factor of κ^2.

Another point that should be taken into account is that the effective mass m of an electron moving in the crystal differs from the mass m_o of a free electron; in semiconductors and metals m is normally less than m_o. This leads to the binding energy decreasing by a factor of m/m_o. Thus the ionization energy of an impurity electron in the crystal, $\mathcal{E}_i$, differs from that of a hydrogen atom, $\mathcal{E}_H$, by a factor of $m/m_o\kappa^2$, i.e., $\mathcal{E}_i=(\mathcal{E}_H/\kappa^2)\,m/m_o$.

This result may be arrived at in a somewhat more rigorous fashion. If an impurity generates, for an electron in the crystal, an extra potential $V(\vec{r}) = -e^2/\kappa r$ which varies comparatively smoothly (varies little when separated by a distance equal to the lattice spacing), it is possible to reduce the Schrödinger Equation (1.2) with this potential to the form

$$\left(- \frac{\hbar}{2m}\nabla^2 - \frac{e^2}{\kappa r} \right) \psi(\vec{r}) = \mathcal{E}\,\psi(\vec{r}) . \tag{1.23}$$

For a smooth potential $V(\vec{r})$ the Equation (1.2) therefore reduces to the effective mass Equation (1.23), where the periodic potential $U(\vec{r})$ is absent and the free-electron mass is replaced by the effective mass m. The Equation (1.23) coincides with the Schrödinger equation for a hydrogen atom if the effective mass m is replaced by the mass m_o and if the quantity $e/\kappa^{1/2}$ is replaced by the charge e. The energy values that have been found from the Equation (1.23) differ from the energies for a hydrogen atom, like the equations themselves. If we take the conduction band edge to be zero energy, the bound states of an electron on a donor will lie below the band edge:

$$\mathcal{E}_{in} = \frac{me^4}{2\kappa^2 \hbar^2} \frac{1}{n^2} \quad (n = 1,2,\ldots) \quad . \tag{1.24}$$

For impurity levels formed by V and III elements in Ge with $m = 0.1m_o$ and $\kappa = 12$, the impurity level ionization energy $\mathcal{E}_{i1}$ is equal to 0.01 eV. Since $\mathcal{E}_{i1} \ll \mathcal{E}_g$ (for Ge $\mathcal{E}_g \simeq 0.7$ eV), such levels are called shallow. For the hydrogen-like shallow levels of interest, we may introduce the effective Bohr radius

$$a_B = a_B^0 \kappa \frac{m_o}{m} , \tag{1.25}$$

where $a_B^0 = 0.53$Å is the radius of the first Bohr orbit in a hydrogen atom; this radius corresponds to the wavefunction maximum (the prohability of detecting an electron at r=a is the largest). For Ge $a_B \approx 10^2 a_B^0$.

As the temperature is increased, the concentration of electrons excited from donors into the conduction band (or holes excited from acceptors into the valence band) heightens; accordingly, the conductivity increases, too. At temperatures T 100 > K shallow donors (acceptors) in Ge become completely ionized (are said to be depleted) and the conduction electron (hole) concentration remains unchanged as the temperature is increased until intrinsic conductivity sets in. A compensation effect arises when donors and acceptors concur in a sample. If, for instance, the donor concentration N_D is larger than the acceptor concentration N_A, part of the electrons excited from donors, N_A, will fill the lower-lying acceptor levels and the maximum number of electrons which the donor can supply to the conduction band will be $n=N_D - N_A$. For $N_D > N_A$, when conduction is realized by electrons, the semiconductor is said to be n-type; for $N_A > N_D$, when conduction occurs by holes, the semiconductor is said to be p-type (n for negative, p for positive).

The effective mass technique (replacing the Equation (1.2) with the impurity potential $V(\vec{r})$ by the Equation (1.23) is unfit to describe the state of electrons coupled to impurities which form deep levels with $\mathcal{E}_{i1} \sim \mathcal{E}_g$. The lattice spacing a in this case is already insufficiently small compared to (at best several times smaller than) the Bohr radius a_B. It is only with the severe inequality $a_B \gg a$ that one may reduce the Schrödinger equation to the effective mass equation and make use of the dielectric constant κ .

It will be observed that at low temperatures, when the number of phonons is small yet, charged donors and acceptors, which disturb the ideal lattice periodicity, make a dominant contribution to electron and hole scattering.

The foregoing examples of substitutional impurities in Ge, Si, etc. are not the only types of donors and acceptors. Some impurities that lodge in interstices are apt to be donors as well. An example is the Li atom, which has one valence electron. This unpaired electron can readily be removed, i.e., transferred to the conduction band. In respect of the contribution to conductivity, an interstitial donor is similar to a substitutional donor (Sb). Interstitial acceptors have not been detected apparently because of the large radius of the negative ion.

Similar considerations concerning the formation of donors and acepptors hold for semiconducting compounds composed of groups III and V or II and VI elements. A group VI element, as it replaces, for example, an Sb atom in InSb, behaves like a donor, whereas a group II element replacing In behaves like an acceptor. In compounds, donors and acceptors often arise because of the breaking of stoichiometry which may be due to the absence of the atom of one of the constituents in the lattice site (a vacancy) or come from the absence of the extra atom of one of the constituents which is located in the interstice.

For example, a donor center may form if one or several electrons are captured by an anionic vacancy. Classical examples of centers of this type are F centers in alkaline-haloid compounds. If, say, several anion sites are vacant in CdS (which corresponds to a Cd^{2+} ion excess), Two electrons per S^{2-} vacancy should arise in the lattice to maintain the electro-neutrality of the crystal. In the vicinity of a vacant site (vacancy) the charge is positive and the excess electrons will be attracted and captured by the anionic vacancy. It is this entity that will act as a donor. If, in some way or other, one (or both) of these electrons is freed from the trap, it will get into the conduction band; as a result the conductivity of the crystal will increase.

In semiconductors, by contrast with metals, the conductivity in the low-temperature range enhances with increase in temperature because the impurity centers are ionized and the charge carrier concentration increases. In the region of temperatures where all impurities are ionized, the conductivity either is constant or decreases with increasing temperature, just as in metals, because of a decrease in charge carrier mobility which is due to an increase of the probability of charge carrier scattering on phonons. As the temperature is increased further, electrons and holes originate, the extrinsic conductivity region is succeeded by an intrinsic conductivity region, and the electrical conductivity increases exponentially with temperature. Another way of exciting charge carriers into the conduction band or from the valence band is by irradiating the semiconductor by light of specified frequency ω so that $h\omega \geq \mathcal{E}_i$ (extrinsic photoconductivity) or $h\omega \geq \mathcal{E}_g$ (intrinsic photoconductivity).

From what has been stated above, it should be clear that in order to describe conductivity and many other phenomena and properties of solids, it is necessary to know the electron population of states, i.e., the distribution of electrons in states. Electrons obey the Fermi-Dirac quantum statistics which is valid for an ensemble of identical particles with a half-integral spin. Pauli's exclusion principle, according to which not more than one electron may be in each state, holds for such particles (if the spin quantum number is not included into the state characteristic, two electrons with opposite spins may be present in a state). The Fermi-Dirac distribution function has the form

$$f(\mathcal{E}) = \left[1 + \exp\left(\frac{\mathcal{E} - \mathcal{E}_F}{k_B T}\right)\right]^{-1} , \tag{1.26}$$

with $\mathcal{E}_F$ being the Fermi energy. The function $f(\mathcal{E})$ is the occupation probability for a state of energy $\mathcal{E}$. At T=0 K f=1 for $\mathcal{E} < \mathcal{E}_F$ and f=0 for $\mathcal{E} > \mathcal{E}_F$, i.e., the $f(\mathcal{E})$ function has the shape of a step: When $\mathcal{E} = \mathcal{E}_F$, it varies stepwise from 1 to 0. We see that for

T = 0 K all the states of an energy less than the Fermi energy are occupied, the states of higher energy being empty; the Fermi energy is a boundary between occupied and empty states. At temperature T > 0 K, which, however, are low enough, $k_BT \ll \mathcal{E}_F$, we have, as before, f = 1 for $\mathcal{E} \ll \mathcal{E}_F$ and f = 0 for $\mathcal{E} \gg \mathcal{E}_F$; but in the vicinity of $\mathcal{E} = \mathcal{E}_F$ the function $f(\mathcal{E})$ non varies more smoothly rather than stepwise. This smoothing of $f(\mathcal{E})$ occurs over an interval of energies of the order of k_BT in the neighborhood of $\mathcal{E}_F$. For T > 0 K the Fermi level occupation probability f is equal to one half. Let it be noted that at T > 0 K the number of states occupied above the Fermi level ($\mathcal{E} > \mathcal{E}_F$) is exactly equal to the number of unoccupied states below the Fermi level.

For the "tail" of the states that are smeared around the step $f(\mathcal{E})$ when $\mathcal{E} = \mathcal{E}_F$, the distribution function (1.26) reduces to

$$f(\mathcal{E}) = \exp \frac{\mathcal{E}_F - \mathcal{E}}{k_BT} \quad (\mathcal{E} - \mathcal{E}_F \gg k_BT) \quad , \tag{1.27a}$$

$$f(\mathcal{E}) = 1 - \exp \frac{\mathcal{E} - \mathcal{E}_F}{k_BT} \quad (\mathcal{E}_F - \mathcal{E} \gg k_BT) \quad . \tag{1.27b}$$

The functions (1.27) are classical Maxwell-Boltzmann distribution functions. The approximations (1.27) often work well for semiconductors. When $|\mathcal{E} - \mathcal{E}_F| = 4k_BT$, the error arising from use of these formulas amounts to approximately 2%. the electron gas obeying the Fermi-Dirac statistics is said to be degenerate. But if the Maxwell-Boltzmann statistics is valid, the electron gas is said to be nondegenerate.

For any electron distribution in states, the Fermi energy is determined from the electroneutrality equation of the crystal as a whole, i.e., from the requirement that the net positive charge be equal to the net negative charge. If both donors and acceptors are present in the crystal, the neutrality equation is of the form

$$N_D^+ + p = N_A^- + n \tag{1.28}$$

where N_D^+ and N_A^- are the concentrations of charged donors and acceptors, p and n the concentrations of holes and electrons. We multiply the density of states in the conduction band, $\rho(\mathcal{E})$ (see the Equation (1.15)), by the occupation probability $f(\mathcal{E})$ for a state of energy $\mathcal{E}$ and integrate the occupied states of the band with respect to energy. This evidently yields the electron concentration

$$n = \int f(\mathcal{E}) \rho(\mathcal{E}) d\mathcal{E} \quad . \tag{1.29}$$

The hole concentration p may be obtained in a similar fashion, but the function $f(\mathcal{E})$ in the Equation (1.29) should be replaced by $1 - f(\mathcal{E})$, viz., the probability that the valence band state of energy $\mathcal{E}$ is free from an electron (i.e., is occupied by a hole).

In calculating N_D^+ and N_A^-, account must be taken of the spin degeneracy factor β of the impurity level. Assume that the donor is monovalent, i.e., has one electron on the outermost s-shell. Since the electron is an unpaired one, it can be captured by a donor in the crystal in two ways: either with the spin down or with the spin up. Thus two states with different spins correspond to the same electron energy - the state is doubly degenerate, i.e., $\beta = 2$. If the donor level is split off by an n-fold degenerate band from a band that has several extrema, the factor β will be larger than 2. The charged donor concentration, say $N_D^+ = N_D f$, with N being the total donor concentration and f a Fermi distribution function in which the spin degeneracy of the donor has been taken into account, will be

$$N_D^+ = N_D \left(1 + \frac{1}{\beta} \exp \frac{\mathcal{E}_D - \mathcal{E}_F}{k_B T} \right)^{-1} . \tag{1.30}$$

For an acceptor capable of capturing a second electron to complete the outermost shell, the factor $1/\beta$ involved in the Equation (1.30) is 2, for the absence of a second electron may be described in two ways, as a function of spin direction. For an acceptor (e.g., In) in Ge, where the valence band at $\vec{k} = 0$ is four-fold degenerate with respect to spin, $1/\beta = 4$. Knowing from experiment N_D and N_A, and also the density of states $\rho(\mathcal{E})$, which is determined by the nature of the semiconductor, it is possible from the Equation (1.28) to calculate the only unknown quantity, viz., the Fermi energy $\mathcal{E}_F$, and then the electron and hole concentration.

If only one type of charge carriers contributes to transport phenomena, the charge carriers concentration can be determined from a number of experiments. To this end one most frequently makes use of the Hall effect, which consists in the occurence of an emf in a strip of metal placed in mutually perpendicular electric and magnetic fields. If a primary current of density i is aligned with the x axis, then, due to the deflection of electrons by a magnetic field of intensity parallel to the z axis in the y direction, an electric field $E^{\cdot} = RiH$ arises, where the Hall coefficient

$$R = 1/en \tag{1.31}$$

is negative for electrons and $R = 1/ep$ is positive for holes (as a function of the sign of the charge e). On measuring i, H, and E_y, the electron and hole concentration can be calculated in a straightforward manner. This simple relation between R and n(p) is valid for a degenerate charge carrier gas. If the gas is nondegenerate, then $R = A/en$, where the coefficient A is of the order of unity and depends on the mechanism of charge carrier scattering and on the dispersion relation. Since the conductivity due to electrons,

$$\sigma = en\mu_e , \tag{1.32}$$

the product $|R|\sigma$ is equal to the electron mobility

$$\mu_e = |R|\sigma . \tag{1.33}$$

The same thing certainly holds true for holes.

Chapter 2

BAND STRUCTURE OF GAPLESS SEMICONDUCTORS

2.1. The Detection of the Gapless State

As we have seen in Chapter 1, the division of solids into metals and dielectrics is based on the electron energy spectrum exhibiting different structure. What is essential here is the electron occupancy of the energy bands. In dielectrics, part of the bands are wholly occupied and higher-energy bands are free from electrons at temperature T = 0 K. In metals, one or several bands are partially filled with electrons and the forbidden-energy interval is absent.

Quite recently, the division of solids into metals and dielectrics has been found not to be absolute. Of fundamental importance has been the discovery of materials in which the stationary state is intermediate between the metallic and the dielectric state (and, accordingly, the properties of the new materials are intermediate between those of metals and dielectrics). In these materials there is no energy threshold of passage of electrons from occupied to empty states, while the density of the electron gas at T = 0 K is equal to zero.

Such a material may be viewed as a semiconductor devoid of a forbidden energy band or as a metal destitute of free electrons at T = 0 K. It may also be regarded on the one hand, as a semiconductor having an empty conduction band and a completely occupied valence band contiguous to the conduction band and, on the other, as a metal with a half-filled band, but with the density of states equal to zero at the Fermi energy. These materials have received the name of gapless semiconductors (GS).

In 1934-1935 Shubin and Vonsovsky (1934) predicted, on the basis of their many-electron polar model of the crystal, that dielectrics with a continuous energy spectrum may, in principle, exist which have no gap. A substantial contribution to the development of the theory of this peculiar state, which subsequently came to be called the gapless state, has been made by Abrikosov and Beneslavsky (1970), Abrikosov (1974,1975),Gelmont and Dyskonov (1972), and Gelmont et al. (1976a).

The gapless state was first discovered experimentally by Tsidilkovski (1954, 1957) in mercury telluride (HgTe) in 1955-1957. Prior to those investigations, HgTe, just as HgSe, had been considered to be a metal. That conclusion had been based on electrical conductivity measurements according to which the temperature coefficient of electrical conductivity was negative, as should be the case with metals. Using the technique of zone melting followed by ingot annealing, Tsidilkovski (1955) obtained crystals, including single crystals, with an impurity content ($1x10^{17}cm^{-3}$) a factor of 50-100 lower than that in the HgTe samples studied earlier and with record-breaking (for that time) electron mobilities (at 200-250 K the mobility was 1.8 x $10^4cm^2v^{-1}s^{-1}$). That enable two important factors to be established: (1) in the 200-250 K range the conductivity is intrinsic and (2) the energy gap between the conduction band and the valence band is anomalously small.

The conclusion as to the conductivity being intrinsic was drawn on the strength of

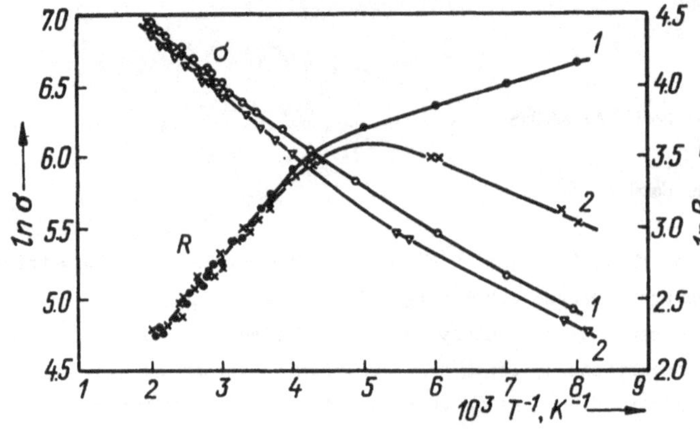

Fig.4. Dependences of Hall coefficient $R(cm^3C^{-1})$ and electrical conductivity $\sigma(\Omega^{-1}cm^{-1})$ on inverse temperature for two HgTe samples (1 and 2) in the 120–525 K range

the fact that the curves describing the temperature dependence of the electrical conductivity, σT, and of the Hall coefficient, R(T), practically merge (Fig.4). This means that individual properties of samples (for example those containing impurities, inhomogeneities, etc.) play no role and the transport phenomena are determined by the electron spectrum of the crystal. The coincidence of the R(T) and σT curves for different samples is the most important signpost to the existence of intrinsic conductivity. The electron and hole concentrations in the intrinsic conductivity region of the semiconductor, n_i, can be determined with the aid of the electroneutrality equation (see Eq.(1.28))

$$n_i = n = p \ , \tag{2.1}$$

where n and p are the electron and hole concentrations.

When the electrons and holes obey the Boltzmann statistics and the energy bands are parabolic (which is practically always true in the immediate vicinity of the band edges), the electron and hole concentrations are given by the expressions (the conduction band edge is taken as zero energy, $\mathcal{E} = 0$)

$$n = N_e \exp\left(\frac{\mathcal{E}_F}{k_B T}\right) \ , \quad p = N_h \exp\left(-\frac{\mathcal{E}_g + \mathcal{E}_F}{k_B T}\right) \ , \tag{2.2}$$

with $N_{e,h} = (1/4\pi^{3/2})(2m_{e,h}k_BT / \hbar^2)^{3/2}$, $m_{e,h}$ the electron or hole effective mass.

By substituting the expressions (2.2) into the Equation (2.1), one can find T dependence of the Fermi energy $\mathcal{E}_F$. Using this dependence, the Equation (2.2) yields the intrinsic charge carrier concentration

$$n_i = 2(m_e m_h)^{3/4} \left(\frac{k_B T}{2\pi\hbar^2}\right)^{3/2} \exp\left(-\frac{\mathcal{E}_g}{2k_B T}\right) . \tag{2.3}$$

Let us now see in what fashion n_i can be found from experimental data. Since, as has been established, the electron mobility in HgTe is much greater than the hole mobility, $\mu_e \gg \mu_h$, the Hall coefficient R_i in the region of intrinsic conductivity will, with good accuracy, be

$$R_i \approx \frac{1}{en_i} \tag{2.4}$$

irrespective of wether the magnetic fields are strong (i.e., $\mu_e H>1, \mu_h H>1$, with H being the magnetic field intensity: the quantity μH indicates to what extent an electron or hole rotates in a magnetic field, to be more exact, it is the angle of rotation tangent) or weak ($\mu_e H<1, \mu_h H<1$). Figure 4 demonstrates that at T>200 K the concentration n_i is larger than the electron concentration $n \approx 1/eR$ found from the Hall coefficient R in the extrinsic conductivity region T>200 K.

On determining the lower bound of the intrinsic-conductivity temperature interval, we will try, by using the formulas (2.3) and (2.4), to determine the value of the energy gap between the conduction and the valence bands. As is manifest from the Equations (2.3) and (2.4), the gap $\mathcal{E}_g$ can be found most conveniently from the $\ln(R_i T^{3/2})$ vs. 1/T curve, which is a straight line. The $\mathcal{E}_g$ value thus calculated from the slope of the straight line has turned out to be unusually small: Within the measurement accuracy it does not exceed ≈ 0.02 eV. The same value of $\mathcal{E}_g$ is obtained from electrical conductivity σ and thermomagnetic Nernst-Ettinghausen effect measurements. To find the value of $\mathcal{E}_g$ from the electrical conductivity, it is necessary to construct a $\ln(\sigma/T)$ vs. 1/T curve, which is a straight line, and to calculate $\mathcal{E}_g$ from the slope of that line. The choice of the function σ/T is dictated by the fact that for the greater part of the intrinsic-conductivity temperature range the electron mobility $\mu_e \sim T^{-1/2}$ and, consequently, $\sigma = en\mu_e \sim T\exp(-\mathcal{E}_g/2k_B T)$.

In 1955-1956 Tsidilkovski measured the angular dependence of magnetoresistance on single-crystalline HgTe, i.e., the variation of the resistivity in a magnetic field $(\rho_H - \rho_o)/\rho_o$ as a function of the angle θ between the electric current and the magnetic field. His measurements showed this dependence to be described by a curve with a maximum located at $\theta = \pi/2$. For $\theta = 0$ and π magnetoresistance is practically equal to zero. This means that the electron dispersion relation is isotropic or, what amounts to the same thing, the isoenergetic surfaces have spherical shape. Therefore, the lowest energy minimum of the conduction band lies at the center of the Brillouin zone, $\vec{k} = 0$. Since the valence band edge of all crystals with zincblende structure lies at $\vec{k} = 0$, one could draw the conclusion that the conduction and valence band edges in HgTe correspond to the same wavevector $\vec{k} = 0$, i.e., that the energy gap between the bands is a direct one.

The energy gap $\mathcal{E}_g$ value found for HgTe turned out to be anomalously small compared with those for all the semiconductors that were known. The value of $\mathcal{E}_g$ is comparatively small in the semiconductor InSb which, as a matter of fact, possesses the same crystallographic structure as that of HgTe, viz., zincblende structure. For InSb the value of the direct gap at $\vec{k} = 0$ constitutes 0.24 eV at 4.2 K, i.e., is an order of magnitude higher than the $\mathcal{E}_g$ value found for HgTe. However, greater attention attaches to another fact.

The value $\mathcal{E}_g = 0.02$ eV, as can be readily seen, is less than the mean thermal energy of charge carriers, k_BT, in the entire intrinsic conductivity region: 0.022 eV at T = 250 K and 0.046 eV at T = 525 K. The quantity k_BT specifies the thermal smearing of the energy levels and, in particular, of the band edges of interest. Since the gap value estimated from experimental data over the interval 200 K $\leq$ T $\leq$ 525 K is less than the thermal energy of electrons, i.e., the band edge smearing, the value $\mathcal{E}_g = 0.02$ eV indicates only the upper bound of the gap value: If a gap is present, it may be only less than 0.02 eV. In principle, the results of these measurements permit also the conclusion that $\mathcal{E}_g = 0$. It will be observed that the formula (2.3) itself is valid for n_i at $\mathcal{E}_g > k_BT$, when the electrons and holes obey the Boltzmann statistics. For the temperature range T>200 K, for which $\mathcal{E}_g < k_BT$, the $\mathcal{E}_g$ values calculated with the aid of the Equation (2.3) should therefore be regarded as a very crude estimate. What remains correct is only the conclusion that the gap should be less than 0.02 eV.

Since there is every reason to believe that the gap $\mathcal{E}_g$ is close to zero, one can try to plot the $R_i(T)$ relation to a log-log scale $\ln R_i(\ln T)$, in order to ascertain whether this dependence obeys the power law $R_i \sim T^n$ (as already stated, one may not set $\mathcal{E}_g = 0$ in the formula (2.3) to obtain $R_i \sim T^{-3/2}$). Figure 5 presents this dependence for two HgTe samples (see Fig.4). As can be seen, above 200 K we have, with good accuracy, $R_i \sim T^{3/2}$ in the entire region of intrinsic conductivity, i.e., $n_i \sim T^{3/2}$. At T<200 K the curves for different samples do not coincide, a fact which indicates different impurity content of the samples. It turns out that even if allowance is made for the nonparabolicity of the conduction band in HgTe (i.e., for the departure of the dispersion relation from the quadratic law $\mathcal{E} = \hbar^2k^2/2m$) and for the comparatively weak electron gas degeneracy, the power law $n_i \sim T^{3/2}$ remains practically invariable. As will be shown later in the text, this dependece of the intrinsic charge carrier concentration is a characteristic feature of gap-

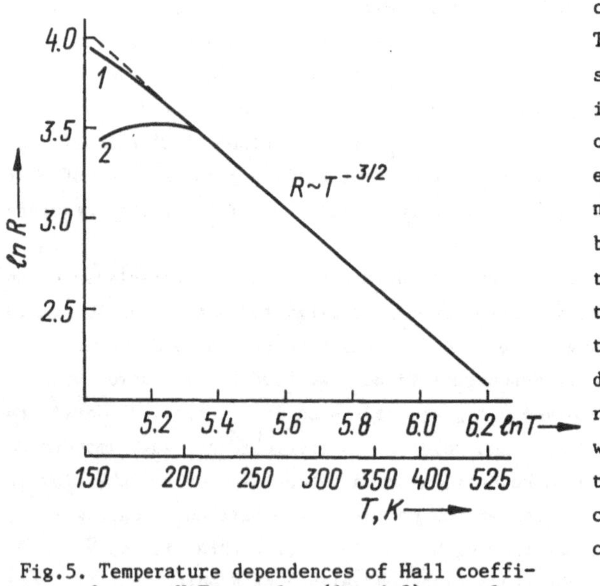

Fig.5. Temperature dependences of Hall coefficient for two HgTe samples (1 and 2) on a log-log scale

less semiconductors, in which the energy gap between the conduction band and the valence band is identically equal to zero. Typically, normal semiconductors exhibit an exponential dependence $n_i(T)$.

Thus pioneering investigations into galvano- and thermomagnetic properties of mercury telluride showed the conduction and valence band extrema to lie at the same value of the wavevector $\vec{k} = 0$ and the value of the direct gap between the bands $\mathcal{E}_g$ to be $\lessapprox 0.02$ eV. However, an unpleasant little cloud soon appeared on the horizon. It is known that the smaller the direct gap between the bands is, the smaller the effective mass of the electron near the band edge, m_e, is. The value of m_e can be estimated using the formula (1.17):

$$\frac{1}{m_e} = \frac{1}{m_o} \left(1 + \frac{\mathcal{E}_p}{\mathcal{E}_g} \right) . \tag{2.5}$$

The quantity $\mathcal{E}_g$ in this equation stands for values of the direct gap between band edges whose wavefunctions possess different symmetry, for example s and p symmetry. It has been found experimentally that for diamond- and zincblende-structure crystals $\mathcal{E}_p \approx 20$ eV (to within 20%). For Ge, Si, and InSb the formula (2.5) leads to reasonable m_e values which agree with experimental data. For InSb, say with $\mathcal{E}_g = 0.24$ eV, the estimate (2.5) yields $m_e = 0.112\ m_o$, which coincides within the experimental error with the experimental value $m_e = 0.014\ m_o$. For Ge (the direct gap at $\vec{k} = 0$ is $\mathcal{E}_g = 0.09$ eV), according to the Equation (2.5), $m_e = 0.043\ m_o$, and the experimental value $m_e = 0.041\ m_o$. If we make a similar estimate for HgTe at the value $\mathcal{E}_g = 0.02$ eV, the Equation (2.5) gives $m_e = 1 \times 10^{-3} m_o$.[5)] However, the findings available in 1955–1956, particularly the thermal e.m.f. value, indicated that the experimental value of m_e exceeded by more than an order of magnitude the quantity estimated with the help of the Equation (2.5) for $\mathcal{E}_g = 0.02$ eV. At the same time, the quantity $m_e \simeq 0.015\ m_o$ found according to the Equation (2.5) for the gap value $\mathcal{E}_g \simeq 0.3$ eV, which corresponds to a peak in the light absorption spectrum, was in reasonable agreement with experimental data for m_e. The impression was conveyed that two valence bands exist: One of them is separated from the conduction band edge by $\mathcal{E}_g < 0.02$ eV (and possibly touches the conduction band at $\vec{k} = 0$) and does not affect the electron effective mass value; the other, which determines the value of m_e, is separated from the conduction band by $\mathcal{E}_g \simeq 0.3$ eV.

A final understanding of the band structure of mercury telluride was gained only after similar and other difficulties in explaining experimental facts for gray tin had been surmounted with the aid of what is known as the inverse band model, proposed by Groves and Paul in 1963 (1963, 1964). Therefore we start by focusing our attention on gray tin.

5) In 1955–1956 the author estimated the value of m according to a formula that is valid in the nearly free electron approximation:

$$\frac{1}{m_e} = \frac{1}{m_o} \left(1 + \frac{8\pi^2\hbar^2}{m_o a^2 \mathcal{E}_g} \right) .$$

with a being the lattice spacing. The m_e values thus calculated are fairly close to those given by the formula (2.5).

2.2. Gray Tin

2.2.1. Crystal Structure

The Periodic Table group IV element No 50 - tin - exists in two modifications, viz., white (β-Sn) and gray (α-Sn). If stored lengthily (from several days to several years) at a temperature below +13.2°C, the usual white metallic tin turns spontaneously into nonmetallic gray tin. The transformation accelerates considerable if α-Sn nuclei are present. So one tin object may infect another. From the earliest times people have known what is called the "tin plague", which infected, for example, tin buttons of army full-dress uniforms stored in unheated storehouses. The transformation of white tin to gray tin is accompanied by a large increase in volume, by 27%, so that a white tin object dissolves into powder. This, by the way, accounts for the difficulties experienced in obtaining gray tin single crystals. The reverse transformation proceeds comparatively rapidly once the temperature rises avobe +13°C. The white tin to gray tin transformation is normally accelerated by inoculating (crystallizing) gray tin from powder, by subjecting white tin to cold working or by soaking it at a temperature of about -30°C.

Gray tin is always obtained as a powder if the transformation of lumps of white tin is effected without taking special precautions. It is known that studying the electrical conductivity, galvanomagnetic, and other phenomena does not permit one to extract information concerning the energy spectrum and the mechanism of charge carrier scattering in a material; to this end single crystals are required. Although the low-temperature phase of tin had been discovered as far back as 1851 (O. Erdman), it is only in 1958 that it became possible to grow first gray-tin single crystals (Ewald and Tufte, 1958). Crystals were grown from a liquid phase (amalgam) and contained therefore considerable less defects than samples prepared when solid tin underwent transformation. The crystals were grown by continuous slow crystallization (the rate of crystal growth was approximately equal to ≈ 1 cm per month; a slow rate of growth is known to promote the formation of crystals that are more perfect) from a saturated solution in mercury at a temperature between -20 and -30°C. The best of the crystals thus grown, more than 1 cm long, had the same purity as that of the initial tin.

Gray tin has the same crystal structure as the group IV elements-carbon (in the structure of diamond), silicon, and germanium (the lattice of white tin is not a cubic one, as in gray tin, but a tetrahedral one; the unit cell is a rectangular parallelepiped with a square at the base). In a diamond-like crystal lattice it is possible to single out a group of eighteen atoms which form a cubic cell (Fig.6). To imagine the arrangement of the atoms in this group, we break up a cube with centered faces and edge length a into eight equal cubes with a/2 edges. There are atoms at the centers of four of these cubes. One of these cubes is indicated by a dotted line in the left-hand upper front corner of Figure 6. Each atom in the diamond lattice, as is seen from Figure 6, is located at the center of a regular tetrahedron and is surrounded by four atoms which lie at the vertices of the tetrahedron. Forces of directional covalent bonds are at work between these atoms in the tetrahedron. It is simple matter to see that of the eighteen atoms pictured in Figure 6, as many as eight atoms belong to a cubic cell of

Fig.6. Structure of diamond-like crystals. Clear and dark spheres denote atoms of different face-centered sublattices

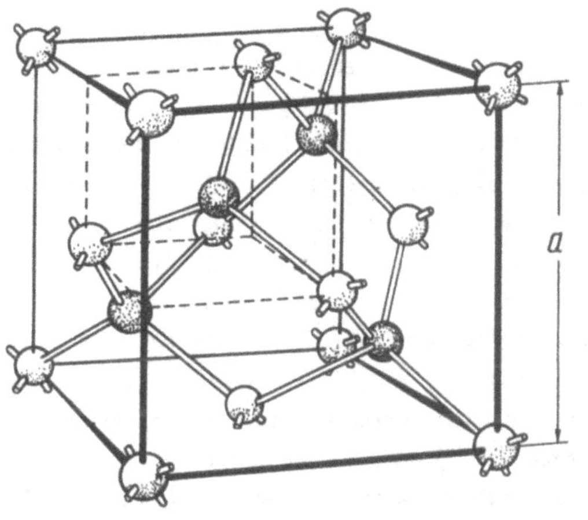

volume a^3. Indeed, of the eight atoms at the cube vertices there is one atom per cube, since each vertex is common for the eight cubes. Of the six atoms at the centers of the faces there are three atoms per cube, since each face is common for two cubes. Within a cube there are four atoms. Thus 1 + 3 + 4 atoms pertain to a cube of volume a^3.

The diamond lattice can be viewed as consisting of two identical face-centered lattices displaced relative to each other along the body diagonal of the cube by one fourth of the length. Each atom of one sublattice (say, dark spheres in Figure 6) is tetrahedrally surrounded by four atoms of the other sublattice (clear spheres).

The electron energy spectrum, i.e., the arrangement and shape of the energy bands, is determined by crystal symmetry, the properties of the atoms that form the crystal, and by the character of their bonds between each other. From a knowledge of these three components, one can ascertain not only the qualitative picture of the band structure of a material but also quantitatively calculate it with sufficient accuracy. The confidence of the band scheme is verified by experimental studies of galvanomagnetic, optical, and other phenomena.

In a free tin atom the valence electrons in the ground (nonexcited) state have the configuration $5s^2 5p^2$. In a gray-tin crystal, as in carbon, the valence electrons in the ground state possess a configuration such as sp^3: $5s5p^3$. Each atom in the lattice with four valence electrons forms four equivalent sp^3-bonds which are directed toward the tetrahedron vertices (Fig.6). By use of the wavefunctions describing these covalent bonds, it is possible to construct the wavefunctions of electrons in a crystal to a one-electron approximation and, in principle, to calculate the band structure with the help of the techniques developed. For lack of an opportunity to outline this procedure here, we shall go another way.

2.2.2. Original Band Schemes

We will try to recapitulate the major features peculiar to the band model of gray tin on the basis of an analysis of known experimental findings. In the review published in 1960, Bush and Kern (1960) considered all the known experimental results of measurements of photoconductivity, light absorption, magentic susceptibility, and of a number of transport phenomena (electrical conductivity, Hall effect, magnetoresistance, and thermal e.m.f.). These authors arrived at the conclusion that gray tin is a semiconductor with a rather small energy gap: $\mathcal{E}_g = 0.09$ eV. However, the photoconductivity data, indicating a maximum near 0.08 eV, proved to be irreproducible: Other maesurements gave the value $\mathcal{E}_g = 0.22$ eV. In the infrared range between 0.03 and 0.06 eV, no light transmission was observed at liquid-nitrogen or liquid-helium temperatures. In analyzing transport phenomena, one encountered the usual difficulties which came from the fact that either there was no information at all about the values and temperature dependences of electron and hole mobilities, about the degree of degeneracy of the electron gas, and even about the number of energy bands contributing to charge transfer, or the evidence available was very unreliable.

The most trustworthy of all the experimental data was the value of activation energy which is obtained from the temperature dependence of the electrical conductivity in the region of highest temperatures (in the vicinity of $0^{o}C$), where the intrinsic conductivity $\sigma \sim \exp(-\mathcal{E}_g/2k_BT)$ was assumed to exist. According to these measurements, the value of $\mathcal{E}_g$ ranges between 0.08 and 0.09 eV. However, the aforementioned uncertainties did not permit one to answer the questions whether $\mathcal{E}_g$ is an energy gap between the conduction band and the valence band and, if this is a gap between the bands, whether it is a direct one or an indirect one . The contingency of this being the impurity activation energy could not be dismissed, either. If, for example, $\mathcal{E}_g = 0.09$ eV were taken to be a direct gap at 0 K, then, on the assumption that the gap variation temperature coefficient $d\mathcal{E}_g/dT$ is approximately the same as that in other group IV semiconductors, the value obtained for the temperature 273 K would be $\mathcal{E}_g \simeq 0$. In that case the character of the transport phenomena should be different from that observed.

For a long time the band structure of gray tin appeared to be enigmatic. The versions of band schemes that were assumed were incapable of accounting for the entire set of experimental data. Only Groves' and Paul's inverse band model (1963) has allowed an understanding to be grained of the experimental material known. In their analysis, Groves and Paul proceeded from a number of reliably established experimental facts which earlier could not be explained from a unified point of view but seemed to be at variance.

It has been found that the magnetoresistance of α-Sn samples exhibiting n-type electronic conductivity is anisotropic, i.e., depends on the magnetic field orientation relative to the crystallographic axes. However, as the temperature is lowered below 273 K the anisotropy decreases, vanishing altogether at 77 K. The character of the anisotropy indicates that the lowest conduction band minima in the Brillouin zone lie in the <111> directions. The disappearance of the anisotropy may be explained in two ways. With decreasing temperature the electron scattering on impurity ions, which is more anisotropic than the vibrational scattering that prevails at high temperatures, gains in importance and may therefore compensate for the electron effective mass anisotropy in the magnetoresistance. Generally speaking, such complete compensation appears to be improbable. More verisimilar is the assumption that as the temperature is lowered the electrons move from the <111> symmetry conduction band minimum to the lower-lying minimum $\vec{k} = 0$.

Studies of Shubnikov-de-Haas magnetoresistance oscillations[6)] on α-Sn samples show that the oscillation period does not depend on the magnetic field orientation relative to the crystallographic axes. From this fact, it follows that the electron dispersion relation $\mathcal{E}(\vec{k})$ is isotropic (in the case of an anisotropic dispersion relation the oscillation periods should be different for the various magnetic field directions). And this, in turn, is possible only provided that the conduction band minimum is located at the center of the Brillouin zone $\vec{k} = 0$.

From the temperature dependence of Shubnikov-de-Haas oscillation amplitudes, the electron effective mass was determined. The electron effective mass m_e turned out to be equal to 0.02 m_0, which is approximately a factor of 5 higher than the value calculated according to formula (2.5) for $\mathcal{E}_g = 0.08$ eV on the assumption that the energy $\mathcal{E}_p = 20$ eV, just as for other group IV semiconductors and group III and V element compounds.

There is one more inconsistence of experimental evidence for gray tin with the band model of a direct-gap semiconductor. With so narrow an energy gap (0.08 eV), the conduction band at $\vec{k} = 0$ should be substantially nonparabolic, i.e., as the wavevector increases the dispersion relation $\mathcal{E}(\vec{k})$ should deviate increasingly from the quadratic law. The electron effective mass in this case should increase with energy, i.e., as the conduction band is filled with electrons. At the same time, measurements of Shubnikov-de-Haas oscillations have shown the electron mass to be independent of electron concentration, i.e., conduction band occupancy.

2.2.3. Inverse Band Model

The band scheme proposed by Groves and Paul is based, on the one hand, on comparing interband energy gaps at $\vec{k} = 0$ in the isoelectronic series of group IV semiconductors and type III-V and II-VI compounds and, on the other, on analyzing the influence of hydrostatic pressure on the Hall effect and electrical conductivity of gray tin. According to Herman's perturbation method (see, e.g., Tsidilkovski, 1982, and also Sect. 2.3.1.), the energy gap in the isoelectronic sequencies is $\mathcal{E}_g \equiv \mathcal{E}(s)-\mathcal{E}(p) \doteq \mathcal{E}_g(IV)-b\lambda^2$, where $\mathcal{E}(s)$ and $\mathcal{E}(p)$ are the energies of s- and p-symmetry band edges, λ is a measure of the antisymmetric part of the crystalline potential, and b is a constant. For group IV

6) Whenever an electron energy level (Landau level) coincides with the Fermi level when the magnetic field is varied, a peak arises on the magnetoresistence as a function of field, i.e., the magnetoresistance oscillates. This is what is called Shubnikov-de-Haas oscillations. The oscillation amplitude depends on temperature and electron effective mass. By measuring the temperature dependence of oscillation amplitudes the effective mass can be determined. From the period of oscillations, the electron concentration can be found.

elements with a symmetric potential $\lambda = 0$, for III-V compounds $\lambda = 1$, and for II-VI compounds $\lambda = 2$.

Figure 7 presents the dependence of $\mathcal{E}(s)-\mathcal{E}(p)$ on λ^2 for two isoelectronic series. With the CdTe-InSb straight line linearly extrapolated to gray tin, the energy gap turns out to be equal to -0.2 eV. But if one extrapolates a line of the same curvature as that in the germanium series (from GaAs to Ge), the result for $\mathcal{E}_g$ will be -0.4 eV. Thus the λ^2 dependence of $\mathcal{E}_g$ indicates that the energy gap $\mathcal{E}_g$ is negative,

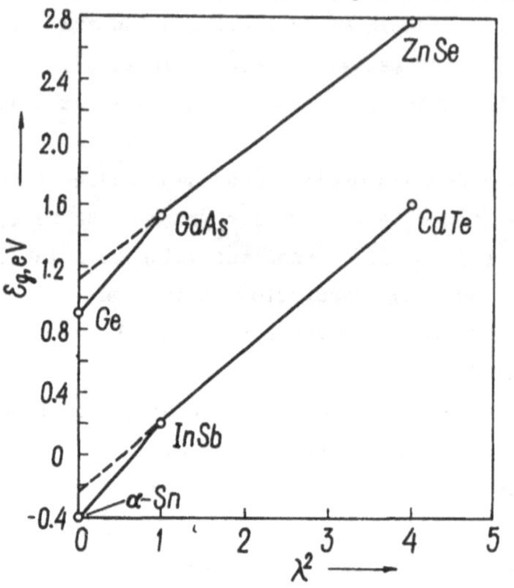

Fig.7. Dependence of the energy gap $\mathcal{E}_g = \mathcal{E}(s)-\mathcal{E}(p)$ on the parameter λ^2 for isoelectronic sequences of Ge and α-Sn. The dashed lines indicate segments where the curves are extrapolated to $\lambda^2 = 0$.

Fig.8. Band diagram of a Ge-type semiconductor with a gap (a) and of GS α-Sn (b) with inverted band order

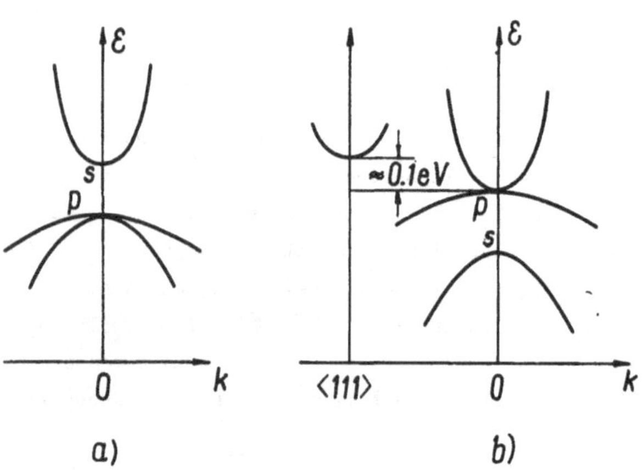

i.e., the p-symmetry band lies higher in energy than the s-symmetry band.

Put another way, the s- and p-bands of gray tin are arranged in the reverse order as compared with germanium. Figure 8 depicts the relative position of the conduction and valence bands for Ge and α-Sn at $\vec{k} = 0$. The fundamental difference between the band pictures of Ge and α-Sn shown in Figures 8a) and 8b) lies in the fact that in germanium the top of the valence band and the bottom of the conduction band are separated by an energy gap, whereas in gray tin the conduction band minimum and the valence band maximum coincide (the bands are degenerate at $\vec{k} = 0$), i.e., the energy gap between these bands is identically equal to zero.

The energy band curvature, proportional to the second-order derivative $d^2\mathcal{E}/dk^2$, is evidently inversely proportional to the effective mass $m = \hbar^2/(d^2\mathcal{E}/dk^2)$. The curvature of the α-Sn light-hole s-band (corresponds to the conduction band of Ge) and the p-conduction band(corresponds to the light-hole band in Ge) is specified chiefly by the quantities $\mathcal{E}_p$ and $\mathcal{E}_g$ (see Eq.(2.5)). For this reason, the curvature of these bands in Ge and α-Sn may be of different sign according to the sign of the gap $\mathcal{E}_g$ ($\mathcal{E}_g$ is a positive quantity). The curvature of the valence p-band in α-Sn, which touches the conduction band (corresponds to the heavy-hole band in Ge) at $\vec{k} = 0$, is determined, just as in Ge, by the energy separation from the bands lying higher in energy and thus remains invariable as the s- and p-bands are inverted (transition from Ge to α-Sn).

It may now be readily verified that the formula (2.5) permits one to obtain the correct (i.e., experimentally established) value of the electron effective mass m_e = 0.02 m_o for the values $\mathcal{E}_g$ = -0.4 eV and $\mathcal{E}_p$ = 30 eV (the latter value differs somewhat, but within reasonable limits, from the mean value $\mathcal{E}_p \approx 20$ eV for other crystals). Using the quoted values of $\mathcal{E}_g$ and $\mathcal{E}_p$ as well as the energy separation from remote bands, determined in light reflection experiments, one can find the effective masses of light holes, m_{lh} = 0.06 m_o, and heavy holes, m_h = 0.03 m_o. It is also clear that the departure of the conduction band from parabolic shape should, in keeping with experiment, be inappreciable since the energy gap $\mathcal{E}_g$ responsible for the degree of nonparabolicity is large. This agrees with the experimental fact that the electron effective mass is independent of electron concentration, i.e., band occupancy.

The band scheme of α-Sn, portrayed in Figure 8b), permits also explaining results of transport phenomena, photoconductivity, and light reflection measurements. To start with, one had to understand the origin of the activation energy of electrical conductivity, approximately equal to 0.08 eV, since the direct gap between the conduction band and the valence band is equal to zero. It was suggested that this activation energy is due to the indirect gap and that the corresponding conduction band minimum lies at the wavevector $\vec{k} = \pi/a$ (111). In other words, it is assumed that the conduction band minimum of <111> symmetry is separated from the heavy-hole band top of <000> symmetry by about 0.08 eV. Our objective now is to evaluate the relative contribution which the electrons from the <111> and <000> minima make to the transport phenomena and other effects when the temperature is elevated.

The effective mass of <111> minimum electrons can be calculated using the known values of the energy gaps at $\vec{k} = \pi/a$ (111) in Ge and α-Sn and from the effective masses in Ge at the same value of $\vec{k}$. If, in addition, one assumes the energy gap $\mathcal{E}(111)-\mathcal{E}(000)$ to depend on temperature, just as in the affinitive InSb crystal (for which this dependence has been established) and uses donor concentration values found from the Hall effect at low temperatures, he will be able, at last, to calculate the electron concentration in the <111> and <000> minima and the heavy-hole concentration near the band edge <000>.

In order to calculate the contributions of charge carriers of different types to the conductivity $\sigma = en\mu$, it is necessary to know, in addition to the concentrations n, the charge carrier mobilities μ . The latter are much more difficult to calculate than the concentrations are. This is so because a knowledge of the partial scattering probabilities is required. Therefore one has to confine himself to estimates made by extrapolating relevant data for Ge and InSb. One also makes use of the electron and hole effective masses found for α-Sn. Such estimates show that the contribution of <111> minimum electrons to the total conductivity at $T \approx 273$ K does not, at any rate, exceed 10%.

The peculiarities in a number of physical properties of gray tin now become clear. The activation energy of electrical conductivity and the increase in photoconductivity at light quantum energies close to 0.08 eV are due to excitation (by heat or by light) of electrons from the valence band <000> into the conduction band minimum <111>. The occurence of a magnetoresistance anisotropy as the temperature is raised is also associated with an electronic transition into a <111> type minimum, where the dispersion relation is anisotropic (the isoenergetic surfaces have the shape of ellipsoids rather than spheres, as is the case at $\vec{k} = 0$), and the magnetoresistance should therefore be anisotropic.

It becomes also clear why the experimental electron effective mass values are comparatively large at high temperatures: The electrons are located in <111> type minima, which are four in number (there is only one <000> minimum), so the number of states per unit energy interval is greater than the number of states in the <100> minimum, and the effective mass proportional to the number of states is higher. The inverse band scheme clarifies also why no light transmission occurs in α-Sn at helium temperatures: In sufficiently pure samples the absorption edge due to transitions of electrons from the valence band into the conduction band corresponds to an energy that is equal to zero.

2.2.4. Experimental Confirmations of the Inverse Model

The inverse band structure of α-Sn was corroborated by many experimental findings. Very conclusive evidence is provided by experiments aimed at studying the effect of hydrostatic pressure on the physical properties of α-Sn. Two proofs of the band picture of α-Sn have been obtained in these experiments (see Fig.8b)). It has been found experimentally that under hydrostatic pressure $\mathcal{P}$ conduction band minima of the same symmetry in group IV semiconductors and III-V compounds move relative to the top of the heavy-hole band practically at the same rate; put another way, the pressure coefficients $d\mathcal{E}_g/d\mathcal{P}$ are nearly equal in magnitude. Thus the pressure coefficients for all these crystals are

$$\frac{d}{d\mathcal{P}} \mid \mathcal{E}_c(000) - \mathcal{E}_v(000) \mid \approx 12 \times 10^{-6} \text{ eV/bar} ,$$

$$\frac{d}{d\mathcal{P}} \mid \mathcal{E}_c(111) - \mathcal{E}_v(000) \mid \approx 5 \times 10^{-6} \text{ eV/bar}, \tag{2.6}$$

where $\mathcal{E}_c$ and $\mathcal{E}_v$ label the energies of the conduction and valence band edges respectively.

Measurements of the pressure effect on conductivity near 273 K, where the conductivity σ increases exponentially with temperature: $\sigma \sim \exp(-\mathcal{E}_g/2k_BT)$, have shown that the pressure coefficient $-2k_BT \; d(\ln\sigma)/d\mathcal{P} = d(\mathcal{E}_g)/d\mathcal{P}$ of the energy gap $\mathcal{E}_g$ in α-Sn is approximately equal to 5×10^{-6} eV/bar. This was one of the most important indications that a fundamental contribution to conductivity is made by the electrons belonging to the <111> minima. The result obtained applies to the <111> minimum of the conduction band, but it does not contradict the conclusion that the minimum $\mathcal{E}_c(000)$ lies lower in energy than the minimum $\mathcal{E}_c(111)$, since the contribution to conductivity near 273 K by <000> minimum electrons is sufficiently small and the variation of this contribution with pressure introduces only a small correction.

The second proof supporting the inverse band structure of gray tin, obtained in hydrostatic pressure experiments, is based on measurements of he temperature dependence of the Hall coefficient R(T). It has turned out that for the purest n-type samples over the 20-100 K range $R \sim T^{-3/2}$. Since the electron effective mass is much less (approximately a factor of 10 lower) than the heavy-hole mass, the electron mobility should considerably exceed the hole mobility. Therefore $R \simeq 1/en$. Consequently, the electron concentration behaves as $n \sim T^{3/2}$ (Fig.9). Per se (without allowing for the effect of pressure), this dependence over a large temperature interval for pure samples (i.e., samples with a relatively small impurity content) is a serious demonstration that the energy gap is equal to zero. If the gap were finite, the n(T) relation would be an exponential one. Let us see in what fashion the $n \sim T^{3/2}$ law obtains in the case where the gap between the conduction and valence bands is equal to zero.

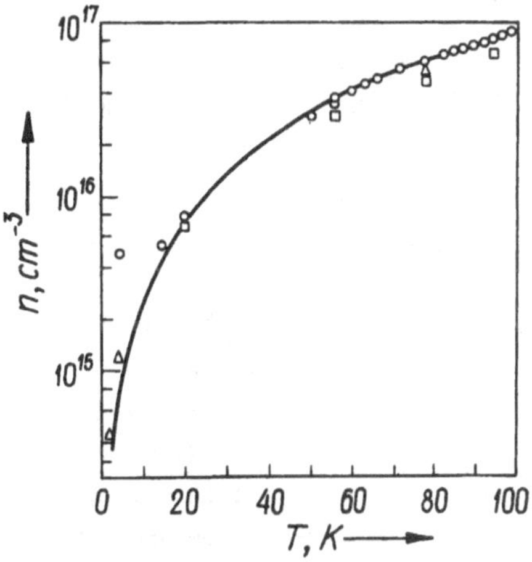

Fig.9. T dependence of electron concentration $n = 1/eR$ (R is the Hall coefficient) for an n-type α-Sn sample with $N_D = 5 \times 10^{15} cm^{-3}$. The solid line refers to the dependence $n \sim T^{3/2}$. 1 stands for data obtained at atmospheric pressure, 2 for data at pressure 2 kbar, 3 for a purer sample at atmospheric pressure.

Imagine a perfectly pure (impurity-free) zero-gap semiconductor (see Fig. 8b)). At T=0 K the electrons wholly fill the valence band, whereas no electrons are present in the conduction band. Evidently, the Fermi level will pass through the degeneracy point of two (valence and conduction) p-bands. It is a simple matter to see that the temperature is raised, when electrons appear in the conduction band and holes appear in the valence band, the Fermi level will necessarily emerge in one of these two bands. This is a very important feature of GS. In which of the bands the Fermi level will emerge may be ascertained from a knowledge of the ratio between the electron effective mass m_e and the hole effective mass m_h. In α-Sn, just as in other gapless semiconductors, $m_e < m_h$. Hence it follows that the Fermi level should be in the conduction band.[7)]

7) Indeed, the Fermi energy $\mathcal{E}_F$, say, for a quadratic dispersion relation (this restriction is immaterial as the conclusion is valid for an arbitrary dispersion relation) is equal to $\hbar^2 k_F^2/2m$, with k_F being the wavevector at the Fermi level. It is clear that when the effective mass is smaller the Fermi energy is larger. Therefore, at $m_e < m_h$ the Fermi level will lie in the conduction band.

In such a situation the electron gas will be degenerate (i.e., will obey the Fermi-Dirac statistics), and the hole gas nondegenerate (i.e., will obey the Boltzmann statistics). The electron concentration n and the hole concentration p are related to the Fermi energy and temperature by the following formulas

$$n = \frac{(2m_e \mathcal{E}_F)^{3/2}}{3\pi^2 \hbar^3} , \quad p = 2\left(\frac{m_h k_B T}{2\pi \hbar^2}\right)^{3/2} \exp\left(- \frac{\mathcal{E}_F}{k_B T} \right) . \qquad (2.7)$$

Equating the electron and hole concentrations, i.e., assuming the conductivity to be intrinsic, we find the Fermi energy as a function of temperature[8]:

$$\mathcal{E}_F = k_B T \ln \left[\frac{3\pi^{1/2}}{4} \left(\frac{m_h}{m_e}\right)^{3/2}\right] . \qquad (2.8)$$

Thus $\mathcal{E}_F \sim T$ and, according to the Equation (2.7), $n \sim T^{3/2}$. It was found experimentally that the purer the α-Sn sample, the lower the temperatures where this law holds. At very low temperatures extrinsic charge carriers begin to play a significant role. From the fact that above the temperature $T_{min} = 20$ K the concentration $n \sim T^{3/2}$, it follows that the energy gap between the bands, if it exists, is at any rate less than $k_B T \approx 0.002$ eV. However, other experimental findings, particularly hydrostatic pressure investigations of the Hall effect, indicate that the gap should be identically equal to zero in keeping with the inverse band scheme.

Hall effect measurements in α-Sn under hydrostatic pressure $\mathcal{P} = 2$ kbar over the interval 20–100 K have shown that the $n \sim T^{3/2}$ relation does not alter and, therefore, the gap between the conduction band and the valence band remains equal to zero. If α-Sn had a band structure such as that of Ge (see Fig.8a)), then, as has already been stated, the departure of the electron dispersion relation from quadraticity would, owing to the small value of the energy gap between the s- and p-bands, turn out to be appreciable and the n(T) dependence would be stronger than $n \sim T^{3/2}$. Besides, the distance between the s- and p-bands varies appreciably under hydrostatic pressure: At pressure $\mathcal{P} = 2$ kbar this change is as large as 0.024 eV. In this case an appreciable deviation from the $n \sim T^{3/2}$ relation should be observed. The invariance of this dependence under hydrostatic pressure excludes the band model of a semiconductor with a direct gap between the s- and p-bands (see Fig.8a)) and confirms the inverse band model depicted in Figure 8b) .

Indeed, if the degeneracy of the bands at $\vec{k}=0$ is not accidently but fundamental, i.e., due to crystal symmetry, it should not be perturbed (the gap should remain equal to zero) under hydrostatic pressure, which does not affect this symmetry. We clarify the concepts "accidental band degeneracy" and "fundamental band degeneracy".

It was found that if the total number of electrons of all unit cell atoms is odd, the material is necessarily a metal (Fig.10a)). But if the number of electrons in a cell is even, several different situations are liable to occur (Figs.10b)–e)). In the

8) In the formula (2.8) a term of the order of $k_B T \ln[\ln(m_h/m_e)]$ is omitted which is small compared to the right-hand side of Eq.(2.8).

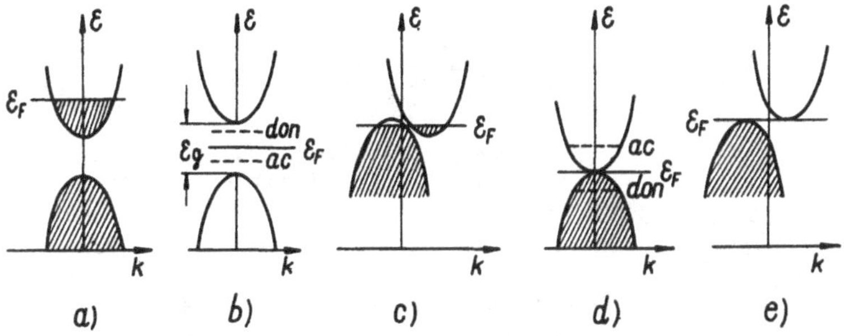

Fig.10. Energy band diagrams for a metal (a), a semiconductor (b), a semimetal (c), a GS in which the band edges lie at the same value of k (d) and at different k (e)

case of a small energy overlap of the conduction and valence bands the substance (e.g., Bi, Sb, As) is a semimetal (Fig.10c)). Because of the overlap, part of the valence band electrons pass over to the conduction band, the boundary energy of the electrons and holes adjoins the free states, and the conductivity of the substance is metallic (no energy is needed to activate charge carriers into free states, as opposed to a dielectric). A semimetal differs from a metal in only that, because of a small degree of overlap, the electron and hole concentration in a semimetal is appreciably less than that in a metal. Normally the electron and hole concentration in a semimetal ranges between 10^{17} and $10^{19} cm^{-3}$.

In GS (Figs.10d), e)) the conduction and valence band edges correspond to the same energy. At T=0 all conduction band states are vacant and all valence band states are occupied. The Fermi level coincides with the energy of the degenerate-band edges. The energy gap may become equal to zero owing to either accidental or fundamental band degeneracy. The accidental degeneracy of bands that touch at the same value of the wavevector $\vec{k}$ (see Fig.10d)) is due to a particular value of the crystalline potential and arises normally between bands of different symmetry, for example between the s- and p-bands. An example of accidental degeneracy is also the situation, depicted in Figure 10e), when the band edges correspond to the same energy but different wavevector $\vec{k}$ values, i.e., possess different symmetries. Accidental degeneracy can be attained at certain values of external isotropic effects: hydrostatic pressure, temperature, and alloying, which do not alter the symmetry of the crystal lattice. The same factors, naturally, can lift the degeneracy and give rise to a gap in the energy spectrum. The gapless state, which results from accidental degeneracy, occurs in the solid solutions $Pb_{1-x}Sn_xTe$, $Pb_{1-x}Sn_xSe$, $Bi\ Sb_{1-x}$ at some ratios of the components, i.e., at some values of x.

Fundamental band degeneracy is due to crystal symmetry and occurs in high-symmetry, chiefly cubic crystals. Within one energy band the electron states differ in wavevectors $\vec{k}$. Each state corresponds to a certain vector $\vec{k}$. Fundamental degeneracy cannot occur at an arbitrary $\vec{k}$. It may arise only for electron wavevectors that are invariant under one or several crystal symmetry operations[9)] (rotation, reflection, etc.). Put another way, fundamental degeneracy arises only at high-symmetry Brillouin zone

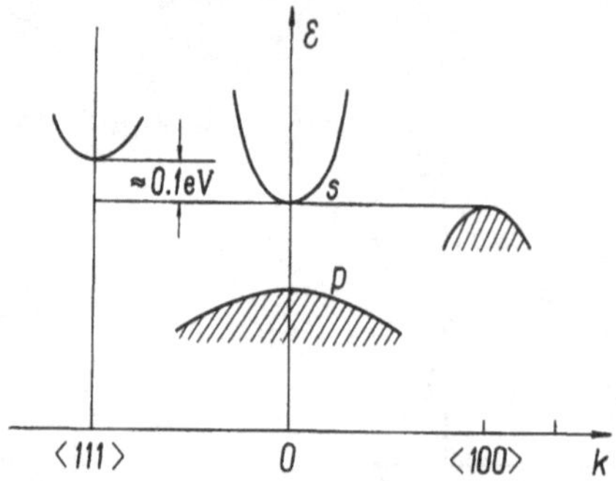

Fig.11. Version of the band model of gray tin with the top of the valence band at $k \neq 0$

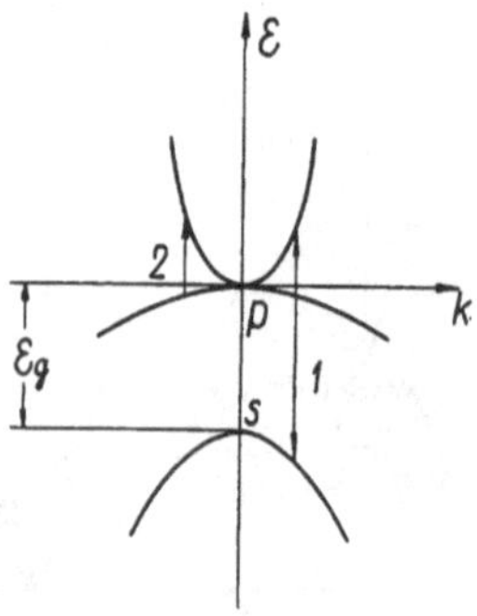

Fig.12. Interband transitions of electrons in α-Sn (1) between s and p bands, (2) between p bands

points, particularly at the point $\vec{k} = 0$ which possesses the highest symmetry. Fundamental degeneracy may be removed only by the action of a perturbation which lowers the crystal symmetry: uniaxial compression or stretching, magnetic field. Under the effect of an isotropic perturbation the gapless state associated with fundamental degeneracy persists. Fundamental conduction and valence band degeneracy occurs in GS HgTe, HgSe, $Hg_{1-x}Cd_x$ Te ($x<0.16$), and other. For α-Sn we could examine also the band model, sketched in Figure 11, which differs from the inverse band model presented in Figure 8b). Here the energy gap between the conduction band and the highest-lying valence band is equal to zero, but the top of this band lies on the |100| axis, i.e., at $\vec{k} \neq 0$. The degeneracy of the bands here is accidental, as in Figure 10e) . The valence band at $\vec{k}=0$ is separated by an energy gap from the conduction band. In what is this model attractive? It permits explanation of some experimental facts that cannot be understood if one proceeds from the band model of a semiconductor with a gap (see Fig.8a)).

Indeed, inspection of Figure 11 shows that in the case of a zero indirect gap between the conduction and valence bands the direct gap between the s- and p-bands at $\vec{k}=0$ may turn out to be large enough for the effective mass of the electrons in the s conduction band to be equal to the experimental value 0.02 m_o. However, the fact that the temperature dependence of electron concentration $n \sim T^{3/2}$ persists also under high hydrostatic pressure, i.e., that the gap between the conduction band and the valence band remains equal to zero, excludes the model portrayed in Figure 11 for α-Sn.

For the gap between the conduction band minimum $\mathcal{E}_c(000)$ and the valence band maximum $\mathcal{E}_v(100)$ to remain constant in this model, the rates of motion of these extrema with pressure, i.e., the pressure coefficients[10] $d\mathcal{E}_c(000)/d\mathcal{P}$ and $d\mathcal{E}_v(100)/d\mathcal{P}$ should be equal in magnitude. But since the pressure coefficient for the extremum $\mathcal{E}_c(000)$ is larger than that for $\mathcal{E}_c(111)$ (see Eq.(2.6)), the energy gap between the bands $\mathcal{E}_c(111)$ and $\mathcal{E}_v(100)$ ought to decrease with increasing pressure. But experiment contradicts this: The intrinsic conductivity at temperatures above 160 K decreases exponentially with pressure, a situation that may occur only when the energy gap increases with pressure.

There is also the following argument to support the inverse model of Figure 8b) and to disprove the models shown in Figures 8a) and 11 for α-Sn. The variation of the elec-

9) As a result of such a symmetry operation, the tip of the wavevector, marked by a dot in the Brillouin zone, passes on to an equivalent position, say from the |100| axis to the |010| axis or from the |111| axis to the |111| axis.

10) These pressure coefficients describe the rate of change of the energies $\mathcal{E}_c(000)$ and $\mathcal{E}_v(100)$ relative to an arbitrarily chosen origin, say relative to the extremum $\mathcal{E}_v(000)$.

tron concentration $n=1/eR$ with pressure can be calculated for all the three models. For the inverse model depicted in Figure 8b) the variation of the concentration is due to a change in effective mass, which can be calculated from the pressure dependence of the energy gap between the s- and p-bands. The results of this calculation are in good agreement with experiment. For the models presented in Figure 8a) and 11, the variations of the electron concentration with pressure result from changes in the corresponding activation energies, i.e., in the gaps between the highest valence band and the conduction band. The variations of n calculated for these turn out to be larger than the observables.

The inverse band structure of α-Sn was confirmed most convincingly by investigations of the variation of the light reflection coefficient in a magnetic field (Groves et al., 1969). In a high magnetic field quasicontinuous bands with respect to energy group into discrete levels - Landau levels. When the light quantum energy becomes equal to the distance between the occupied Landau level of the valence band and the empty Landau level of the conduction band, the photon is absorbed and a transition of the electron from the valence band into the conduction band occurs. In this case a peak is observed in the absorption spectrum. As the magnetic field at which the spacing between the Landau levels changes is varied, more and more electronic transitions fall "in resonance" with the frequency of light, and the magnetic field intensity dependence of the light reflection coefficient will have an oscillatory character. Similar oscillations occur also for the light reflection coefficient.

In investigations of the magnetic field dependence of the reflection coefficient on gray tin, two oscillation series associated with two types of interband transition, 1 and 2, (Fig.12) were observed. Type 1 transitions are due to the excitation by light of electrons from the light-hole s-band into the conduction band, and type 2 transitions come from the excitation of electrons from the heavy-hole p-band. The magnetic field dependence of the photon energies corresponding to the maxima of the reflection coefficient for transitions 1 and 2 are curves which, with the magnetic field intensity H tending to zero, converge into one point for each series. Extrapolation of the series 1 curves to the magnetic field H=0 has led to a photon energy value of 0.41 eV, which is the gap ε_g between the s- and p-bands. Series 2 curves at $H\to 0$ go to a photon energy equal to zero, which corresponds to a zero gap between the conduction band and the highest-lying valence band. This is perhaps the most straightforward evidence for the inverse band structure of gray tin.

By contrast with hydrostatic pressure, uniaxial pressure lowers the crystal symmetry and lifts the degeneracy of p-bands at $\vec{k}=0$. Experiments aimed at measuring the Hall coefficients on pure α-Sn crystals under uniaxial pressure have shown that the temperature dependence of the Hall coefficient becomes exponential (Roman and Ewald, 1972). This corresponds to the appearance of a gap between the previously degenerate conduction and heavy-hole bands,which augments by 10 meV when the pressure is increased by 1 kbar.

In the group of diamond-like semiconductors (C, Si, Ge, α-Sn), the inverse band structure of gray tin is quite legitimate. With increasing atomic number from diamond to gray tin, the gap between the conduction and valence bands at $\vec{k}=0$ diminishes and vanishes for gray tin. The cause for the occurence of an inverse band structure and of a zero energy gap is discussed in the section that follows.

2.3. Mercury Chalcogenides HgTe and HgSe

2.3.1. Crystal Structure and Herman's Perturbation Method

Compounds made up of elements belonging to the third and fifth (III-V) as well as the second and sixth (II-VI) groups of the Periodic Table crystallize in the zincblende (ZnS) structure. The zincblende lattice may, similar to the diamond lattice, be pictured as consisting of two mutually penetrating face-centered cubic lattices. This structure differs from diamond in that the atoms of the two sublattices (see Fig. 6, dark and clear spheres) are different: In and Sb, Hg and Te, etc.

All the characteristics of diamond-structure crystals, which follow from symmetry properties, hold good also for crystals with zincblende structure as long as we neglect the electron spin.

Herman (1955) has noticed that the similarity of the diamond and zincblende structures permits the main feature of the band structure of zincblende-type crystals to be obtained from the familiar model of the corresponding diamond-type crystals. This way of establishing the major features peculiar to the band structure has received the name of Herman's perturbation method.

Every III-V or II-VI compound may be imagined as formed from an isoelectronic IV-IV compound or a corresponding Periodic Table group IV element by replacing two group IV atoms by the neighbouring group III-V or group II-VI atoms. For example, cubic boron nitride BN is referred to diamond, AlP to silicon, GaS and ZnS (as well as AlSb and InP) to germanium, InSb and CdTe to gray tin. Boron phosphide BP and AlN may be referred to SiC. Other III-V and II-VI semiconductors may be referred only to hypothetic IV-IV compounds such as SiGe, GeSn, etc.

Using the perturbation theory method, it is possible to find the band structure of III-V or II-VI compounds if one knows the band model of the corresponding group IV crystal (or IV-IV compound). The crystalline potential $U^{IV}(\vec{r})$ of a diamond-type crystal is invariant under inversion with respect to the middle of the distance between next nearest neighbours,i.e., for example between atoms with the coordinates (000) and a(1/4, 1/4, 1/4) (the second atom is separated from the first atom by a distance of one fourth the length of the body diagonal of an elementary cube, see Fig.6). Put another way, the potential $U^{IV}(\vec{r})$ is symmetric with respect to the permutation of two next nearest atoms in the unit cell: $U^{IV}(\vec{r})=U^{IV}(-\vec{r})$. By contrast with diamond, these atoms in the zincblende structure are different, i.e., the zincblende structure possesses no center of symmetry, so the crystalline potential $U(\vec{r})$ should contain an antisymmetric part $U^{A}(\vec{r})$. In order to go from the higher symmetry of diamond to the lower symmetry of zincblende, one has to add to the symmetric potential U^{IV} some quantity which is proportional to the antisymmetric potential U^{A}. We will regard this addition $\Delta U(\vec{r})$ as a perturbation of the symmetric part of the potential and write it in the form $\Delta U=\lambda U^{A}$, where the parameter λ specifies the degree of perturbation. For simplicity, the parameter is taken to be equal to 1 or 2 for III-V and II-VI compounds respectively. For II-VI crystals the potential $U(\vec{r})$ is represented as

$$U(\vec{r}) = U^{IV}(\vec{r}) + 2U^{A}(\vec{r}) .$$

In the first order of perturbation theory the energy correction due to the antisymmetric part of the potential is equal to zero for all values of the wavevector $\vec{k}$, exept for the Brillouin zone boundary in the <100> direction ($\vec{k}=2\pi/a$ (100)), where the degenerate state of the diamond structure splits up into two states in the zincblende structure. This splitting is proportional to the first-order perturbation λ and, therefore, the further the components of a compound are removed from the group IV elements, the greater this splitting is (for example, it increases in the isoelectronic InSb-CdTe-AgI series with respect to gray tin). For all the other symmetric Brillouin zone points, the energy gaps of a III-V or II-VI compound are expressed in terms of the counter parts of a IV (or IV-IV) crystal using second-order corrections of perturbation theory

$$\mathcal{E}_g = \mathcal{E}_g(\mathrm{IV}) + b\lambda^2 ,$$

with b being some constant.

By use of Herman's perturbation method, band pictures of GaAs and cubic BN were obtained from the band structures of germanium and diamond respectively. This method permits ascertaining only the major features and qualitative peculiarities of the band structure, but it is incapable of yielding as exact quantitative results as those obtained by widely used band structure calculation methods such as the orthogonalized plane waves method, the pseudopotential method, the Green's function method, etc. We are interested in this method because it was instrumental in finding the energy gap between the s- and p-bands at $\vec{k}=0$ (see Sec.2.2.).

2.3.2. Inverse Band Model

We have already mentioned in Section 2.1. that in 1955-1956 an anomalously small direct energy gap between the conduction band and the valence band was discovered in HgTe, which was at any rate less than 0.02 eV, i.e., less than the mean thermal energy of charge carriers at the temperature T_{min} = 200 K bounding from below the region of intrinsic conductivity. In subsequent years (up to 1964) a number of researchers performed electrical conductivity and Hall effect measurements on HgTe, HgSe, and alloys of these compounds. Analysis of the aforementioned measurements led the authors to either the band model of a semiconductor with a direct gap $\mathcal{E}_g \simeq 0.01$ to 0.02 eV (HgTe) or the model of a semimetal with several overlapping conduction and valence bands (HgTe, HgSe, HgTe-HgSe). The latter model is presented in Figure 13. From the high-lying valence band whose maximum, under the assumption, lies at $k\neq 0$ above the conduction band minimum, part of the electrons pass to the conduction band, and the material is a semimetal. The origin of the high-lying va-

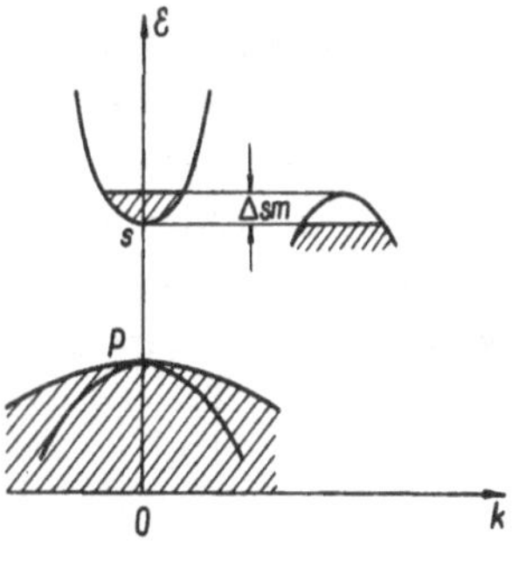

Fig.13. Band model of the semimetal for HgTe, HgSe, and HgTe-HgSe, proposed in 1961-62. Δ_{sm} is the band overlap energy.

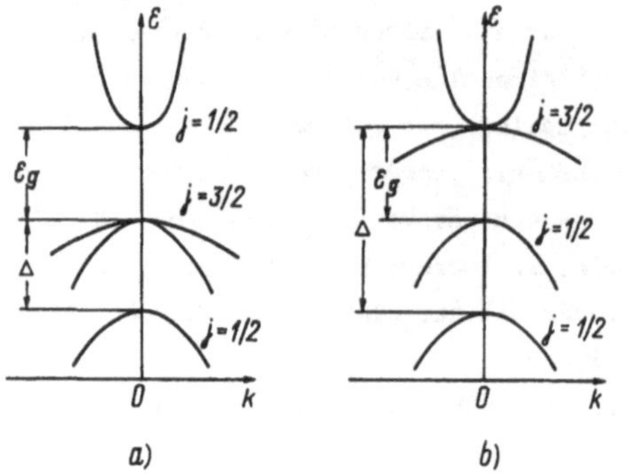

Fig.14. Band diagram of an InSb-type semiconductor (a) and GS HgTe (b) with allowance for the spin-orbit interaction

lence band at $\vec{k}\neq0$ was very difficult to explain since this band disagreed with the band scheme for the familiar III-V semiconductors with zincblende structure. Indeed, for the band sequence typical of III-V semiconductors with the s-band lying above the p-band, there are only two possible explanations of the origin of valence bands whose maxima are located above the bottom of the conduction band: (1) either this is a different valence band, in no way connected with the conduction s-band and the valence p-band, (2) or this is a bend of the heavy-hole p-band by several tenths of an electron volt. Neither of these explanations is, however, verisimilar since the resulting band scheme differs drastically from the band pictures for all the other crystals with zincblende structure.

Only the assumption as to the inverse order of the s- and p-bands, just as in α-Sn, has made it possible to offer a reasonable explanation of the entire body of experimental evidence for HgTe and HgSe. Figure 14, which is similar to Figure 8 for Ge and α-Sn, presents band schemes near $\vec{k}=0$ for an InSb-type semiconductor and an HgTe-type zero-gap semiconductor.

In InSb, the conduction band in the minimum is described by a s-symmetry wavefunction (orbital quantum number $l=0$), and the valence band in the maximum by a p-type symmetry function (orbital quantum number $l=1$). If one does not allow for the electron spin, the conduction band is nondegenerate, and the valence band is triply degenerate. With allowance for the spin, the conduction band would be doubly degenerate and the valence band six-fold degenerate if one disregards the spin-orbit interaction. The latter proves, however, to be very substantial. It leads to partial removal of the degeneracy of the valence band, and the extreme points of the bands are classified according to the total angular momentum $\vec{j} = \vec{l} + \vec{s}$ rather than the orbital angular momentum $\vec{l}$. The conduction band remains doubly degenerate: $j = 0 + 1/2 = 1/2$. For the valence band, the quantum number j may assume two values: $j = 1 + 1/2 = 3/2$ and $j = 1 - 1/2 = 1/2$. The value $j = 3/2$ refers to the light- and heavy-hole bands, each of which is doubly degenerate. The common top of these bands thus is four-fold degenerate. The value $j = 1/2$ refers to the valence band split off from the other two valence bands by the spin-orbit interaction by the amount Δ .

In the inverse band model that holds for HgTe (Fig.14b)), the band of s-symmetry ($j=1/2$) lies below the bands of p-symmetry ($j=3/2$) has a negative curvature, and thus is, in contradistinction to the case of InSb, a valence band. As a result of inversion, the curvature of one of the degenerate bands with $j = 3/2$ turns out to be positive (conduction band), and that of the other negative (heavy-hole band). In this band scheme, the forbidden band, i.e., the gap between the highest valence band and the lowest conduction band, is identically equal to zero, and the energy gap between the extreme points of the s- and p-symmetry bands, $\mathcal{E}_g = \mathcal{E}(s) - \mathcal{E}(p)$, which for InSb is the positive width of the forbidden band, is negative in the inverse band model.

2.3.3. Experimental Evidence for the Inverse Band Model

We now turn to experimental findings which provide evidence that HgTe and HgSe possess an inverse band struture. An important characteristic of GS is the behavior exhibited by the temperature dependence of the intrinsic charge carrier concentration. In contrast to a finite-gap semiconductor where this dependence is activational in character, the curve in gapless semiconductors displays a power-type behavior: $n_i \sim T^{3/2}$ (see Sect. 2.2). In 1955-1956 (Tsidilkovski, 1955,1957) this dependence in HgTe was shown to occur above 200 K. In 1965 it became possible to obtain purer crystals in which the lower boundary of the region of intrinsic conductivity was at $T \approx 60$ K and the dependence observed down to this temperature was $n_i \sim T^{3/2}$.

Since the energy gap between the s- and p-bands in HgTe and HgSe is comparatively small ($\mathcal{E}_g = -0.3$ eV for HgTe and -0.2 eV for HgSe, about the same as that in InSb), the resulting nonparabolicity of the conduction band could be expected to cause substantial departures of the $n_i(T)$ relation from the $n_i \sim T^{3/2}$ law. An analysis shows, however, that in the case of HgTe the nonparabolicity of the conduction band distorts this dependence very little, if at all.

As the conduction band is filled with electrons, i.e., when the electrons are far from the band minimum, the electron effective mass increases, i.e., becomes a function of energy $m_e(\mathcal{E})$ (see Eq.(1.21)). This should lead to an appreciable deviation of the temperature dependence of the Fermi energy from the linear behavior (see Eq.(2.8)), which is typical for parabolic bands, and thus to departure from the $n_i \sim T^{3/2}$ law. In HgTe, just as in other zero-gap semiconductors of this type, the energy gap between the s- and p-bands, on which the electron effective mass depends, varies sensibly with temperature. With increase in temperature $|\mathcal{E}_g|$ decreases and, therefore, the mass m_e (see Eq.(1.11)) decreases. This decrease in mass in HgTe practically compensates for the increase in mass due to conduction band nonparabolicity, and the power law $n_i \sim T^{3/2}$ holds with good accuracy (Harus et al., 1983).

It should be noted that among the GS, HgTe is the most siutable entity for investigating the temperature dependence of the intrinsic charge carrier concentration over a wide temperature range. Thus deviations from the power law $n_i \sim T^{3/2}$ in α-Sn set in at temperatures $T \gtrsim 100$ K owing to thermal excitation of valence band electrons into the conduction band minimum of symmetry <111> which lies approximately 0.08 eV higher than the <000> minimum (see Fig.8b)). HgSe and HgSe based solid solutions always contain many donors. As a result of this, the electron concentration of the conduction band

$n \gtrsim 10^{17} cm^{-3}$, so that it is virtually impossible to observe the region of intrinsic conductivity. For comparison, we will state that in the purest HgTe samples the concentration of electrons excited from donors is about $10^{15} cm^{-3}$.

Weighty supportive evidence for the inverse band model of HgTe and HgSe has been obtained in measurements of the kinetic effects under hydrostatic pressure. Hall effect measurements on pure HgTe samples have shown, just as in the case of α-Sn, that the temperature dependence of the Hall coefficient $R \sim T^{3/2}$ persists under a hydrostatic pressure of 5 kbar over the entire temperature range in which intrinsic conductivity occurs (Porowski and Zakrzewski, 1965). Suchlike behavior of the Hall effect (more precisely, the intrinsic electron concentration) must be expected for the inverse band model portrayed in Figure 14b) since the zero energy gap between the conduction band and the highest-lying valence band is a consequence of crystal symmetry and hydrostatic pressure cannot affect the gap value.

Another proof of the inverse band structure of HgTe is based on an analysis of the displacement of s- and p-symmetry bands under hydrostatic pressure. The pressure coefficients of the energy gaps between the s- and p-states do not depend on the relative energy scale position of these states. They depend only on the symmetry of the states. In the numerous experiments aimed at investigating the influence of hydrostatic pressure on transport and optical phenomena in crystals with diamond and zincblende structure, it has been found that as the pressure is raised the s-symmetry band moves upward in energy relative to the p-symmetry band. So the energy gap between the s- and p-bands $\mathcal{E}_g = \mathcal{E}(s) - \mathcal{E}(p)$ increases under pressure $\mathcal{P}$ $(d\mathcal{E}_g/d\mathcal{P} > 0)$ for the "normal" order of bands, i.e., for semiconductors with a gap $\mathcal{E}_g > 0$ (see Fig.14a)), and decreases $(d|\mathcal{E}_g|/d\mathcal{P} < 0)$ for the inverse order of bands, i.e., for a gapless semiconductor with a negative gap $\mathcal{E}_g < 0$ (Fig.14b)). Thus investigating the effect of hydrostatic pressure allows one to choose between the two possibilities: $\mathcal{E}_g > 0$ or $\mathcal{E}_g < 0$.

An appropriate effect for ascertaining the sign of $\mathcal{E}_g$ is the thermal electromotive force. The thermal e.m.f. coefficient α in the case of a degenerate electron gas (because of small effective mass m_e the electron gas in HgTe and HgSe at T<100 K is degenerate) is defined by the formula

$$\alpha = \frac{\pi^3}{3} A \frac{k_B T}{e} \frac{1}{\mathcal{E}_F} , \qquad (2.9)$$

where A is a coefficient of the order of unity, which depends on the mechanism of electron scattering. Since the Fermi energy $\mathcal{E}_F$ depends on effective mass m_e and electron concentration n (see Eq.(2.7)), one can, by measuring the thermal e.m.f. and the Hall coefficient R=1/en, ascertain how the effective mass varies with pressure.

From the way the effective mass varies with pressure, one can judge, on the basis of the formula (2.5), the variation of the gap $\mathcal{E}_g$. The point here is that the quantity $\mathcal{E}_p$ involved in the formula (2.5) is approximately the same for all crystals with diamond and zincblende structure, i.e., depends weakly on the value of lattice spacing. This allows the energy $\mathcal{E}_p$ to be roughly considered as independent of pressure. In such a case the electron effective mass m_e will vary under hydrostatic pressure as $|\mathcal{E}_g|$: for $d\mathcal{E}_g/d\mathcal{P} > 0$ $(\mathcal{E}_g > 0)$ we shall have $dm_e/d\mathcal{P} > 0$; for $d|\mathcal{E}_g|/d\mathcal{P} < 0$ $(\mathcal{E}_g < 0)$ we shall have $dm_e/d\mathcal{P} < 0$.

In experiments concerned with measurements of the thermal e.m.f. on n-HgTe crystals, Piotrzkowski et al. (1965) have found that the thermal e.m.f. α decreases when the pres-

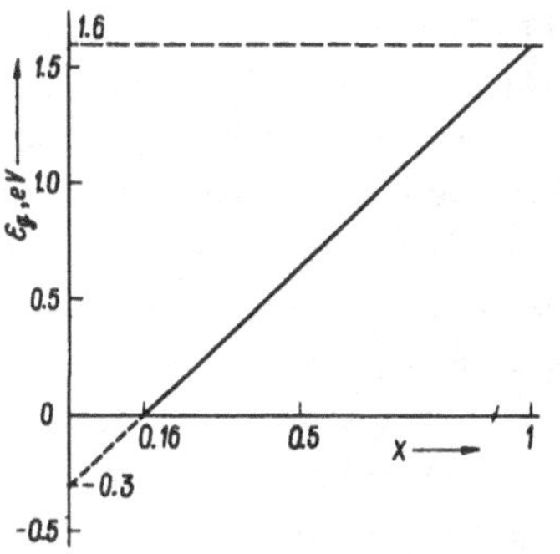

Fig.15. Dependence of the gap $\mathcal{E}_g$ on the Cd content of solid solutions $Hg_{1-x}Cd_xTe$ at T≈4K

sure is increased up to 8 kbar, whereas the Hall coefficient R and, accordingly, the electron concentration remain invariable. Therefore, the decrease of the thermal e.m.f., i.e., the increase in Fermi energy $\mathcal{E}_F$, may be associated only with a decrease in electron effective mass m_e (see Eq.(2.7)). And the decrease of the effective mass with pressure ($dm_e/d\mathcal{P}<0$) indicates that $\mathcal{E}_g<0$, i.e., the band order is inverse (Fig.14b)).

Investigation of the light absorption by HgTe and HgSe crystals have shown the absorption edge corresponding to transitions from the valence band to the conduction band to be displaced by the action of hydrostatic pressure. The pressure coefficient $d\mathcal{E}_g/d\mathcal{P}$ of the energy gap $\mathcal{E}_g=\mathcal{E}(s)-\mathcal{E}(p)$ determined from the absorption edge shift, agrees in magnitude with the values of $d\mathcal{E}_g/d\mathcal{P}$ for the corresponding gaps in other semiconductors with diamond and zincblende structure, and also with the values of $d\mathcal{E}_g/d\mathcal{P}$ obtained in thermal e.m.f. measurements. It may therefore be inferred that the fundamental absorption edge shift testifies in favor of the inverse band model. Investigations of Shubnikov-de-Haas oscillations, magnetophonon resonance[11], light reflection, and helicon wave propagation[12] have corroborated the inverse band order for HgTe. It is of interest to analyze the dependence of the value of the gap $\mathcal{E}_g$ between the s- and p-bands on the molar content of HgTe in $Hg_{1-x}Cd_xTe$ solutions (Fig.15). As the percentage of Cd in a solid solution is decreased, i.e., as x is decreased, the quantity $\mathcal{E}_g$ diminishes monotonically from the value 1.6 eV, which is the energy gap of a "normal" CdTe semiconductor (see Fig.14a)). If HgTe were a semiconductor with a finite gap $\mathcal{E}_g$, the $\mathcal{E}_g(x)$ curve for x varying from 1 to 0 would not intersect the abscissa axis whatever the value of x. However, the experimental dependence $\mathcal{E}_g(x)$ for T=4.2 K crosses the abscissa axis at x≈0.16, and extrapolating it to x=0 yields a negative value, $\mathcal{E}_g=\mathcal{E}(s)-\mathcal{E}(p)= -0.3$ eV for HgTe (see, e.g., Harman et al., 1967).

The most direct and conclusive proofs of the inverse band model of HgTe, just as in the case of α-Sn, have been provided by magnetooptical investigations (Groves et al., 1967, 1970). For HgTe, too, two series of oscillations of the reflection coefficient in a magnetic field were observed which are associated with two types of electronic transition from the light- and heavy-hole bands to the conduction band (see Fig.12). Extrapolation of the magnetic field H dependence of the photon ener-

11) Magnetophonon resonance is a phenomenon that consists in a resonance increase of the probability of inelastic electron scattering by optical phonons under certain magnetic fields. This peculiarity leads to oscillational dependence of kinetic coefficients on magnetic field.

12) A helicon is a weakly decaying low-frequency electromagnetic wave that arises in semiconductors and metals placed in a dc magnetic field H. Helicon waves propagate along the direction of the magnetic field H and are elliptically polarized in the plane perpendicular to H.

gies corresponding to the reflection maxima to a zero field H=0 has led to two values: 0.3 eV and 0 eV. The first photon energy value corresponds to the gap $\mathcal{E}_g$ = -0.3 eV between the light-hole band and the conduction band. The second value indicates that the conduction band and the heavy-hole band touch, i.e., there is no gap between them.

It might be well to point out that the properties of semiconductors in a magnetic field are generally very sensitive to the symmetry of bands whose charge carriers contribute to the phenomenon investigated. Thus in InSb, where the conduction band possesses s-symmetry, the electrons are diamagnetic, whereas in HgTe, where the conduction band has p-symmetry, the electrons are paramagnetic. Magnetic susceptibility studies have revealed the community of the nature of the HgTe conduction band and the InSb and Ge light-hole bands which possess p-symmetry too.

For HgSe crystals the body of direct evidence for an inverse band order is not as large as for HgTe. In some measure this is due to the fact that the HgSe crystals obtained always have a sufficiently large donor content ($\gtrsim 10^{17} cm^{-3}$) and, accordingly, a large electron concentration in the conduction band. However, analysis of the numerous measurements of transport, optical and magnetooptical phenomena permits the indubitable conclusion that HgSe, just as HgTe, possesses an inverse band structure.

2.3.4. The Role of Relativistic Effects

What are the causes that lead to the inverse band structure of the zero-gap semiconductors? If one traces the order in which the bands are arranged at the point $\vec{k}=0$, he can notice that with increase in the atomic number of the crystalforming elements, for example in the group IV element series C-Si-Ge-α-Sn, the direct gap between the bands of s- and p-symmetry decreases. This tendency of the s- and p-bands for converging leads ultimately to their inversion in α-Sn. On the other hand, the role of relativistic effects is known to heighten with increase in the atomic number of an element, i.e., in its charge. Thus it has turned out that it is the relativistic effects which lead not only to a considerable shift but also to alteration of the sequence of levels. For lack of space, these effects, the calculation of which is based on the Dirac relativistic equation [13)], are not treated here; we will only indicate what corrections to the electron energy they generate.

There are three such energy corrections: (1) kinetic energy correction $\mathcal{E}_{mv}$ due to a relativistic increase in electron mass at electron velocities v, when the quantity $(v/c)^2$ becomes appreciable (with c being the velocity of light); (2) correction $\mathcal{E}_D$ to the potential energy of the electron, due to the interaction of the electron with the nucleus - the so-called Darwin correction; (3) correction $\mathcal{E}_{so}$ due to the spin-orbit interaction, i.e., the interaction of the spin magnetic moment of the electron with the magnetic field produced by its orbital motion around the nucleus.

Particularly substantial is the effect which the spin-orbit interaction has on energy level location. The spin-orbit interaction brings about not only a shifting of levels but also a splitting of degenerate states, the splitting being a function of the

13) A relativistic differential equation for the wavefunction of a free particle with spin 1/2, which describes the temporal variation of its state.

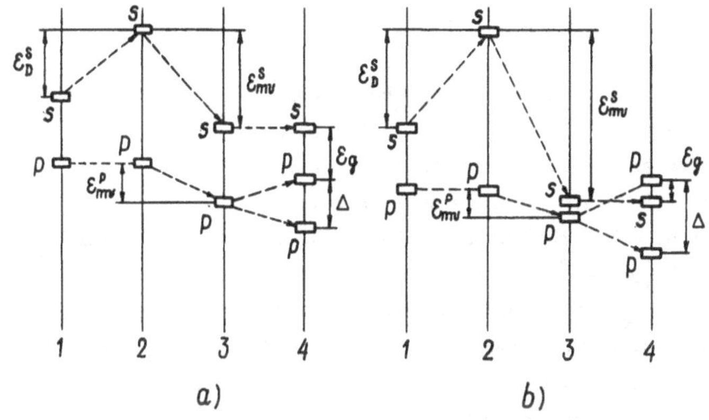

Fig.16. Role of relativistic effects in band structure formation at $\vec{k}=0$ for CdTe(a) and HgTe (b): (1) location of levels without allowance for relativistic corrections; (2) displacement of s symmetry levels by $\mathcal{E}_D^s$; (3) displacement of s and p symmetry levels by $\mathcal{E}_D^s$, $\mathcal{E}_{mv}^s$, and $\mathcal{E}_{mv}^p$; (4) location of levels with allowance for corrections $\mathcal{E}_{mv}$, $\mathcal{E}_D$, and $\mathcal{E}_{so}$

weight of elements - the heavier an element, i.e.,the larger the charge of its nucleus, the stronger the splitting. In contradistinction to spin-orbit interaction, the kinetic and potential energy corrections $\mathcal{E}_{mv}$ and $\mathcal{E}_D$ do not alter the symmetry of the crystalline potential and, therefore, cause no splitting of degenerate levels. These corrections lead only to displacement of levels by the amounts $\mathcal{E}_{mv}$ and $\mathcal{E}_D$, which are much larger for s-symmetry states than for p-symmetry states. The Darwin correction $\mathcal{E}_D$ for p-states is, generally speaking, equal to zero.

The qualitative picture of the influence of relativistic effects on band structure is presented in Figure 16. Shown in the figure is the scheme of band formation at the point $\vec{k}=0$ with account taken of the relativistic effects for CdTe and HgTe in which the lattice spacings differ by less than 0.5%. Consequently, one could expect the values of the gaps between conduction and valence bands to differ little. It is for this reason that the role of relativistic effects in the formation of the band picture can be traced in a more or less pure form. If no allowance is made for these effects, the gap between the s- and p-states in CdTe and HgTe should be about the same. The relativistic corrections $\mathcal{E}_D$ and $\mathcal{E}_{mv}$ for HgTe are much larger than those for CdTe; in consequence, owing to these corrections the gap between the s- and p-states in HgTe is appreciably less than in CdTe. The spin-orbit interaction partially lifts the degeneracy of the p-symmetry band and leads to the four-fold degenerate (with allowance for spin) p-band in HgTe, in contrast to CdTe, lying above the doubly degenerate s-band. Thus it is the relativistic effects that are responsible for the formation of an inverse band structure and, accordingly, of a zero gap between the conduction and valence bands in a gapless semiconductor.

Theoretical band structure calculations from "first principles" [14)] (Herman et al.,1968; Pollak et al., 1970) for α-Sn and HgTe have shown the inverse structure to arise from relativistic effects. Nonrelativistic calculations lead to a "normal" band order for gray tin and mercury chalcogenides, an order that is typical of finite-energy-gap semiconductors.

14) That is to say, nonempirical calculations in which one does not make use of crystal parameters borrowed from experiment. Only parent-atom parameters, symmetry, and interatomic bonds are used.

2.3.5. Shape of the Energy Bands near the Point $\vec{k}=0$

The energy band shape, or, to put it differently, the dependence of the energy of charge carriers $\mathcal{E}$ on their wavevector $\vec{k}$, may, in principle, be calculated on the basis of a Schrödinger equation with an appropriately chosen crystalline potential $U(\vec{r})$. Of greatest interest is usually the shape of bands near extrema whose charge carriers govern transport and a number of other phenomena in semiconductors. In the vicinity of the band extremum located at the point $\vec{k}_o$, the energy may always be expanded in a power series of $k-k_o$, and for small $k-k_o$ one can confine himself to the quadratic expansion term. In the immediate vicinity of the extremum the dispersion relation $\mathcal{E}(\vec{k})$ may therefore be always described with adequate accuracy by a quadratic function, i.e., the band will be parabolic in shape. The constant-energy surfaces in this case will be ellipsoids if $\vec{k}_o \neq 0$; in the case of a cubic crystal they will be spheres if $\vec{k}_o=0$. In going away from the extremum the band shape deviates increasingly from parabolic, the degree of this deviation depending on the proximity of bands that possess a different symmetry. The shape of bands can be studied with the aid of various experimental techniques (galvanomagnetic, resonance, optical and magnetooptical, and other), by varying the charge carrier concentration, i.e., band occupancy.

If the band near the extremum $\vec{k}_o$ is degenerate, the function $\mathcal{E}(\vec{k})$ exhibits a singularity at this point, and the expansion in a power series of $k-k_o$ becomes impossible. In consequence, the isoenergetic surfaces acquire a shape that is more complicated than ellipsoidal or spherical. In order to find the charge carrier dispersion relation for four (allowing for spin) degenerate p-bands at $\vec{k}=0$, it is necessary to write the Hamiltonian $\mathcal{H}$ of the Schrödinger equation. If one confines himself to small values of $\vec{k}$, the problem can be solved with the help of the effective mass method, by retaining terms of order k^2. However, it appears more convenient to obtain the Hamiltonian from symmetry considerations, as was done by J. Luttinger (1956) for a four-fold degenerate valence band.

In considering cubic crystals, we should naturally require that the Hamiltonian be invariant under the symmetry operations of the cube. For a nondegenerate energy band the only invariant that is quadratic in k is k^2. So the Hamiltonian has the form $\mathcal{H} = (\hbar^2/2m)k^2$, where the coefficient m is the effective mass. In the case of nondegenerate bands the state is described not only by the wavevector $\vec{k}$ but also by the total angular momentum vector $\vec{j}$. Since a degenerate state at $\vec{k}=0$ is formed by p-type functions, we are interested in the value $j = 3/2$.

For the state with the quantum number $j = 3/2$ there are in a cubic crystal two more invariants, which are constituted by the products k_α and j_α quadratic in k_α: $(\vec{k}\vec{j})^2$ and $k_x^2j_x^2+k_y^2j_y^2+k_z^2j_z^2$, where j_x, j_y, j_z are numerical 4x4 matrices for a charge carrier with $j = 3/2$. Without writing the Hamiltonian out, we quote the charge carrier dispersion relation which results from it:

$$\mathcal{E}(\vec{k}) = \gamma_1 \frac{\hbar^2k^2}{2m_o} \pm \frac{\hbar^2}{m_o} \left[\gamma_2^2 k^4 + 3(\gamma_3^2 - \gamma_2^2)(k_x^2k_y^2 + k_y^2k_z^2 + k_z^2k_x^2) \right]. \tag{2.10}$$

The dispersion relation (2.10) is specified by the constants γ_1, γ_2, and γ_3. Within the framework of the Luttinger calculation method based on symmetry considerations, the constants γ_1, γ_2, γ_3 are phenomenological parameters whose values must be found from a comparison of computational results with experimental data. In the effective mass method the constants γ_α are expressed in terms of the momentum operator components and the energy gaps between the p-band under consideration and other bands and can, in principle, be calculated.

The isoenergetic surfaces described by the Equation (2.10) are not spherical but possess cubic symmetry. The departure from sphericity is determined by the last term of the Equation (2.10), to be more exact, by the difference of γ_2 from γ_3. The degree of nonsphericity for isoenergetic surfaces of p-bands is therefore specified by the parameter $|(\gamma_2 - \gamma_3)/(\gamma_2 + \gamma_3)|$ which for most of the semiconductors is less than unity and, in order of magnitude, is normally equal to 0.1. In this context one can, for many purposes, confine himself to a spherical approximation, that is, replace the quantities γ_2 and γ_3 by their mean value

$$\bar{\gamma} = \frac{\gamma_2 + \gamma_3}{2}\,. \tag{2.11}$$

In this approximation the two branches of the dispersion relation (2.10) take the form

$$\mathcal{E}_1 = (\gamma_1 + 2\bar{\gamma})\,\frac{\hbar^2 k^2}{2m_o}\,, \quad \mathcal{E}_2 = (\gamma_1 - 2\bar{\gamma})\,\frac{\hbar^2 k^2}{2m_o} \tag{2.12}$$

with the branch $\mathcal{E}_1(\vec{k})$ describing the conduction band, $\mathcal{E}_2(\vec{k})$ the valence band. The electron and heavy-hole effective masses are related to the parameters γ_1 and $\bar{\gamma}$ as

$$m_e = \frac{m_o}{\gamma_1 + 2\bar{\gamma}}\,, \quad m_h = \frac{m_o}{|\gamma_1 - 2\bar{\gamma}|}\,. \tag{2.13}$$

The dispersion relations (2.10) and (1.12) hold only in a small vicinity of the band degeneracy point $\vec{k}=0$, more precisely when the energy $\mathcal{E}(\vec{k})$ is much less than the gap $\mathcal{E}_g$ between the s- and p-bands. Departures from the quadratic dependence $\mathcal{E}(\vec{k})$ (2.12) for electrons are particularly considerable at small gaps $|\mathcal{E}_g|$ (for example, in HgCdTe solid solutions where the gap can be equal to zero). In this case the effective mass approximation is insufficient, and to obtain the electron dispersion relation one should make use of the more general method proposed by Kane (1957, 1979). This method takes into account three closely adjacent bands in the Hamiltonian: the conduction p-band, the light-hole s-band, and the valence p-band split off by spin-orbit interaction (see Fig.14). The other bands are disregarded. The equation to find $\mathcal{E}(\vec{k})$ for the three bands has the form

$$\mathcal{E}(\mathcal{E} - \mathcal{E}_g)(\mathcal{E} + \Delta) - P^2k^2(\mathcal{E} + \tfrac{2}{3}\Delta) = 0, \tag{2.14}$$

where P is the momentum operator matrix element for the conduction p-band and the light-hole s-band, $P^2 = (\hbar^2/2m_o)\mathcal{E}_p$ (see the text preceding the formula (1.17)). The zero of energy is chosen at the degeneracy point of the bands. Three solutions of the cubic Equation (2.14) yield the dispersion relations of the conduction band, $\mathcal{E}_e(k)$ ($\mathcal{E}_e(0)=0$), the light-hole valence band, $\mathcal{E}_{1h}(k)(\mathcal{E}_{1h}(0) = \mathcal{E}_g)$, and the valence band split off by spin-orbit interaction, $\mathcal{E}_{so}(k)$ ($\mathcal{E}_{so}(0) = -\Delta$).

For the heavy-hole valence p-band in the three-band approximation described, the result for the point $\vec{k}=0$ is $\mathcal{E}_h=0$, and the dependence $\mathcal{E}(\vec{k})$ at small k has the form $\mathcal{E}_h(k)=\hbar^2k^2/2m_o$. Here the valence $\mathcal{E}_h$ band has a positive rather than negative curvature, as should be the case with valence bands, and the heavy-hole effective mass m_h is equal to m_o. This means that the three-band approximation does not suffice to describe the heavy-hole p-band. The correct negative curvature and the effective mass m_h consistent with the experimental value are obtained if the calculation allows for the other, remoter bands as well. At the same time the three-band approximation proves acceptable for describing the dispersion relations of conduction-band, light-hole-band, and spin-split-band charge carriers over a sufficiently wide energy range and reflects correctly the gap $\mathcal{E}_g$ and wavevector $\vec{k}$ dependence of the effective masses of the three bands, i.e., the nonparabolicity.

We quote s-band electron dispersion relations $\mathcal{E}(k)$ for some particular cases. If the electron energy $\mathcal{E}$ is much less than the valence band splitting energy Δ (this is always true of HgTe, HgSe, and other zero-gap semiconductors where $\Delta \approx 0.5$ to 1.0 eV and the Fermi energy $\mathcal{E}_F$ is normally less than 0.1 eV), then the cubic Equation (2.14) becomes quadratic

$$\mathcal{E}(\mathcal{E} - \mathcal{E}_g) = \frac{2}{3} P^2k^2. \qquad (2.15)$$

This equation takes into account only two bands: the conduction p-band and the light-hole s-band. The Equation (2.15) yields the electron dispersion relation

$$\mathcal{E}(k) = \frac{\mathcal{E}_g}{2} + \frac{|\mathcal{E}_g|}{2} \left(1 + \frac{8P^2k^2}{3\mathcal{E}_g^2}\right)^{1/2} \qquad (2.16)$$

As is seen, the relation $\mathcal{E}(k)$ may differ substantially from the quadratic dependence. At small $k \to 0$ the root is expanded in a series of k^2 and one obtains a quadratic dispersion relation:

$$\mathcal{E}(k) = \frac{2 P^2}{3|\mathcal{E}_g|} k^2 . \qquad (2.17)$$

If the conduction band and the light-hole band touch when $\vec{k}=0$, i.e., if $\mathcal{E}_g=0$, as is the case, for example, with HgCdTe, then it follows from the Equation (2.14) that for $\mathcal{E} \ll \Delta$ the electron dispersion relation is linear:

$$\mathcal{E}(k) = \left(\frac{2}{3}\right)^{1/2} Pk . \qquad (2.18)$$

For the linear dispersion relation the velocity is $v = \partial\mathcal{E}/\partial k$ = const.

The effective mass in the case of an isotropic nonquadratic dispersion relation, of which the Equations (2.16) and (2.18) are examples, is determined by the first-order energy derivative with respect to the wavevector (1.21).

For the dispersion relation (2.16) which holds in the two-band approximation the electron effective mass is

$$m_e(\mathcal{E}) = m_e^* \left(1 + \frac{2\mathcal{E}}{|\mathcal{E}_g|} \right) , \tag{2.19}$$

where

$$m_e^* = \frac{3\hbar^2 \, |\mathcal{E}_g|}{4P^2} \tag{2.20}$$

is the electron effective mass at the conduction band edge k=0. Thus the larger the gap $|\mathcal{E}_g|$ and the higher the electron energy $\mathcal{E}$, i.e., the fuller the conduction band occupancy, the larger the effective mass m is.

For the linear electron dispersion relation (2.18) the effective mass m_e is a linear function of the wavevector

$$m_e(k) = \left(\frac{3}{2} \right)^{1/2} \frac{\hbar^2}{P} k \quad . \tag{2.21}$$

Effects associated with the nonsphericity of degenerate p-bands, which is reflected in the dispersion relation (2.10) obtained by allowing for the other bands with an accuracy to terms of order k^2 , may manifest themselves in pure samples, when there are few electrons and holes and the band occupancy is inappreciable. But if the samples contain a sufficiently large number of impurities and, consequently, the number of charge carriers is large, a more substantial role is played by effects due to the band nonparabolicity reflected in the dispersion relations resulting from the Equation (2.14). Like the conduction band, the light-hole band is nonparabolic. The spin-split band possesses a lesser degree of nonparabolicity.

2.3.6. Effect of the Electron-Electron Interaction

The energy spectrum that has been described in the previous section is constructed on the basis of the Schrödinger equation with allowance for the symmetry properties of the crystal lattice potential, the latter being taken in the one-electron approximation, when consideration is given to one electron in the averaged field of all the other electrons and all ions. Actually, the zero-gap semiconductor concept itself, as introduced in the foregoing, holds only in the one-electron approximation. However, effects associated with the electron-electron interaction turn out to be very important for GS, and this is what distinguishes them substantially from finite-gap semiconductors.

It is known that owing to the mutual attraction of a hole and an electron a bound state - an exciton - may arise in a semiconductor. If the gap between the bands, $\mathcal{E}_g$, is less than the Coulomb electron and hole attraction energy, i.e., the exciton binding energy $\mathcal{E}_{ex} = \left[m_r e^4/(2\kappa^2\hbar^2) \right] (m_r^{-1} = m_e^{-1} + m_h^{-1})$, the formation of an exciton is energetically more favorable than the excitation of a free-electron -free-hole pair (Knox, 1963). On the other hand, if according to the one-electron treatment the substance should be a semimetal in which the conduction band and the valence band overlap (Fig.10c)), the semimetallic state at a small value of this overlap, $\Delta_{sm} \lesssim \mathcal{E}_{ex}$, turns out to be unstable with respect to electron and hole pairing processes, and the material will be a semiconductor with anarrow energy gap (Keldysh and Kopsev, 1964). The effect of temperature, pressure, and magnetic field may convert such a narrow-gap semiconductor into a semimetal. For a zero-gap semiconductor in which there is no energy threshold for the creation of electron-hole pairs and the energy of the attractive interaction between an electron and a hole is finite, the energy spectrum in the vicinity of the band degeneracy point should evidently be highly distorted by the electron-electron interaction, both the direct Coulomb electrostatic interaction and the exchange interaction.

The energy spectrum of GS, obtained in the one-electron treatment, changes radically if account is taken of the exchange interaction. To clarify the meaning of the exchange interaction, we turn to the simplest example - the hydrogenmolecule. The problem of finding the forces of interaction between four particles (two protons and two electrons) cannot be solved exactly. Therefore one makes recourse to approximate methods, in this case to the perturbation method. It is assumed that in the initial (ground) state the distance R between the atoms is sufficiently large, so that the interatomic interaction may be neglected. With decreasing R the interparticle interaction, which is treated as a small perturbation introducing a correction into the ground state energy of the molecule, enhances. The ground state energy of the molecule is just a sum of the ground state energies $\mathcal{E}_o$ of two hydrogen atoms: $2\mathcal{E}_o = - m_o e^4/\hbar^2 = -e^2/a_B$, where $a_B = \hbar^2/(m_o e^2)$ is the Bohr radius.

The wavefunction of the system of electrons 1 and 2 of two noninteracting atoms a and b is chosen as the product of the wavefunctions of the electrons in the ground 1s-state $\psi = \exp(-r/a_B)$: $\psi(1,2) = \psi_a(1)\,\psi_b(2)$ where 1 and 2 are the coordinates of the electrons $\vec{r}_1$ and $\vec{r}_2$. Since the electrons are fundamentally indistinguishable, one should consider the possibility that electron 1 pertains to atom b, and electron 2 to atom a. The system with the same energy $2\mathcal{E}_o$ is therefore described by the wavefunction $\psi(2,1) = \psi_a(2)\,\psi_b(1)$. Being linear, the Schrödinger equation is satisfied by any linear combination, in particular the sum $\psi_+ = \psi(1,2)+\psi(2,1)$ and the difference $\psi_- = \psi(1,2)-\psi(2,1)$.

The correction W (due to the interatomic interaction) to the system's ground state energy $2\mathcal{E}_o$ is $W = \int V\psi^2 d\vec{r}$, where V is the sum of the Coulomb interaction energies of all electrons and protons, and ψ is ψ_+ or ψ_-. Substituting V and ψ into the integral yields the interaction energy of two hydrogen atoms:

$$\mathcal{E}_+ = 2\mathcal{E}_o + W_+ , \quad \mathcal{E}_- = 2\mathcal{E}_o + W_- ,$$

where

$$W_+ = \frac{B + A}{1 + S^2} , \quad W_- = \frac{B - A}{1 - S^2} ,$$

$$B = \int \psi_a^2(1)\,\psi_b^2(2)\, V d\vec{r}_1 d\vec{r}_2, \quad S^2 = \int \psi_a(1)\,\psi_b(1)\psi_a(2)\,\psi_b(2)\, d\vec{r}_1 d\vec{r}_2,$$

$$A = \int \psi_a(1)\,\psi_b(1)\,\psi_a(2)\,\psi_b(2)\, V d\vec{r}_1 d\vec{r}_2 .$$

The integral S specifies the degree of overlap of the electron wavefunctions (overlap integral), the integral B is the electrostatic interaction energy of two charged clouds (the energy of the interaction of electrons among themselves and with protons) with charge densities $e\psi_a^2$ (1) and $e\psi_b^2$ (2). The quantity A has no graphic interpretation, unlike the direct Coulomb interaction integral B. The integral A, which is called the exchange integral, may be treated as the electrostatic interaction energy of charged clouds with charge densities $e\psi_a(1)\ \psi_b(1)$ (1) and $e\psi_a(2)\psi_b(2)$ (2), which are said to be exchange charge densities. This implies that each of the electrons 1 and 2 is partially in the ground state ψ_a and partially in ψ_b, i.e., they sort of "interchange". The exchange energy is the potential energy at the point $\vec{r}_1$, where the first electron is located, in the exchange density field of the charge localized at the point $\vec{r}_2$, where the second electron is located.

Thus the interaction between the atoms leads to splitting of the nondegenerate ground energy level $2\mathcal{E}_o$ into two levels that differ by the energy 2A. The $\mathcal{E}_-$ vs. R curve which corresponds to the antisymmetric wavefunction ψ_- describes the repulsion of the atoms at arbitrary distances R between them (a monotonically decreasing function). The $\mathcal{E}_+(R)$ curve corresponding to the symmetric function ψ_+ has a minimum at $R_o \approx a_B$: when $R<R_o$ the atoms attract, when $R>R_o$ they repel. The presence of the minimum means that when $R=R_o$ a stable hydrogen molecule forms.

To this point we have disregarded in this section the electron spin and the Pauli exclusion principle. According to the Pauli exclusion principle, the electron system's wavefunction, which depends on both space and spin variables, should be antisymmetric, i.e., should reverse sign when a pair of electrons interchange (in the case of a permutation of four coordinates). If one disregards the interaction of the magnetic moment of the electron with the magnetic field produced by its orbital motion, i.e., neglects the spin-orbit interaction, then, since the electric interaction of particles is independent of their spins (the Hamiltonian of the system contains no spin), the total wavefunction may be represented as the product of the spatial-coordinates wavefunction by the spin-coordinates function. The total wavefunction will be antisymmetric if the symmetric coordinate function ψ_+ is multiplied by the antisymmetric spin wavefunction corresponding to the state with antiparallel spins or if the antisymmetric function ψ_- is multiplied by the symmetric spin wavefunction corresponding to the state with parallel spins. The hydrogen molecule forms only provided the coordinate function is symmetric, i.e., is ψ_+, and, therefore, the electronic spins are antiparallel.

Thus, in spite of the fact that the interaction of electron spins is weak, the energy of the system depends on the total electron spin, as if there were between the electrons an additional, exchange interaction. It is clear that the exchange interaction is part of the Coulomb interaction of electrons. The separation of the exchange interaction results from the electron system being treated approximately; in this treatment the wavefunction of the system is expressed in terms of the counterparts of separate electrons. The exchange interaction is essential only when the electron wavefunctions overlap. If the states ψ_a and ψ_b overlap insignificantly, the exchange integral A is very small.

The exchange energy A, which depends on the coordinates of interacting particles, is a nonlocal potential and may not be introduced directly into the one-electron Schrödinger equation. It turns out that for semiconductors with a full valence band separated from the conduction band by a gap that is wide enough for the interband transitions of electrons to be neglected, the exchange interaction energy may be reduced to some local potential which can be included in the one-electron periodic potential.[15)]Thus the exchange interaction effects are taken into account in the energy spectrum $\mathcal{E}(\vec{k})$ obtained from the one-electron Schrödinger equation. By contrast, the electron exchange interaction energy in metals with partially filled bands may not be reduced in a local potential, since an addition to the one-electron energy arises owing to the exchange interaction; the larger the electron concentration, the larger this addition.

An impurity-free zero-gap semiconductor is similar to a metal with a half-filled band, but the density of states at the Fermi level, in contradistinction to metals, is equal to zero. The exchange interaction between electrons modifies the charge carrier spectrum: Terms appear in the electron and hole dispersion relations (Halperin and Rice, 1968) which are proportional to the first power of the wavevector k (these are first-order terms of perturbation theory):

$$\mathcal{E}_e(k) = \frac{\hbar^2 k^2}{2m_e} + \frac{3\pi e^2}{32\kappa} k , \qquad (2.22)$$

$$\mathcal{E}_h(k) = \frac{\hbar^2 k^2}{2m_h} + \frac{3\pi e^2}{32\kappa} k , \qquad (2.23)$$

where κ is the dielectric constant entering, for example, into the energy of the direct Coulomb interaction of electrons

$$V(r) = - \frac{e^2}{\kappa r} . \qquad (2.24)$$

Formally, it follows from the Equation (2.22) that the exchange interaction makes a substantial contribution to the energy of electrons at wavevectors $k \lesssim a_{eB}^{-1}$, where

$$a_{eB} = \frac{\kappa\hbar^2}{m_e e^2} \qquad (2.25)$$

is the Bohr radius of an electron of effective mass m_e in a crystal with a dielectric constant κ (1.25). Therefore, according to the Equation (2.22), the dispersion relation should be linear in the region of energies $\mathcal{E}$ less than or of the order of the Bohr electron energy $\mathcal{E}_{eB} = m_e e^4/(2\kappa^2\hbar^2)$ (1.24), and for $\mathcal{E} \gg \mathcal{E}_{eB}$ the energy $\mathcal{E}_e(k)$ corrections due to exchange interaction should be small.

15) The exchange interaction may be described by a local potential if the electrons with opposite spins are equal in number, so that the total spin is zero. In crystals with diamond structure two electrons that form a valence bond possess oppositely oriented spins. If part of the electrons are excited into the conduction band, the total spin of the excited electron and of its previous partner in valence bond is no longer equal to zero and the exchange interaction cannot be described with the help of a local potential.

However, the formulas (2.22) and (2.23), obtained within the framework of perturbation theory, hold only for large wavevectors $k \gg a_{eB}^{-1}$, i.e., when the quadratic term in k is larger than the linear term. Hence it follows that in the range of validity of the formula (2.22) the exchange interaction makes a small contribution to the electron dispersion relation.

In GS the heavy-hole effective mass m_h normally exceeds the electron mass m_e by a large amount. In that energy range $\mathcal{E} \lessapprox \mathcal{E}_{eB}$ where the exchange additions to electron energy are, according to the Equation (2.22), large, the Bohr energy $\mathcal{E}_{eB}$ is therefore approximately equal to the exciton binding energy $\mathcal{E}_{ex}$, and many-electron effects become essential. But for holes with $m_h \gg m_e$, the energy corresctions due to exchange interaction are considerable over a wider energy interval than for electrons, up to the Bohr hole energy

$$\mathcal{E}_{hB} = \frac{m_h e^4}{2\kappa^2 \hbar^2} = \mathcal{E}_{eB} \frac{m_h}{m_e} . \tag{2.26}$$

Thus in the energy range $\mathcal{E}_{eB} < \mathcal{E} \leq \mathcal{E}_{hB}$, where the one-electron approximation is applicable, the exchange interaction brings about a substantial readjustment of the valence band spectrum: Along with quadratic terms in the wavevector, linear terms appear in the dispersion relation.

As has already been noted, the influence of many-electron effects near the degeneracy point of the conduction and valence bands at energies $\mathcal{E} \lessapprox \mathcal{E}_{eB}$ cannot be analyzed on the basis of the formulas (2.22) and (2.23) obtained within the framework of perturbation theory. Account must be taken of the so-called dielectric anomaly associated with peculiarities of charge screening in GS. It turns out that the possibility of a spontaneous creation of electron-hole pairs due to the absence of threshold energy for such processes leads to screening of charged centers (e.g., impurity ions) by charge carriers. For the same reason, the expression for the direct Coulomb interaction potential (2.24) with the constant quantity κ proves unsatisfactory. As opposed to the case of a semiconductor with a gap, the dielectric constant κ for the region of energies near the degeneracy point of the bands in zero-gap semiconductors depends sharply on the wavevector $\vec{q}$ and electromagnetic wave frequency ω . The difference of these dependence for gapless and normal semiconductors tells upon directly observed conductivity and light absorption coefficient values. Thus in a semiconductor with a gap $\mathcal{E}_g$ the absorption coefficient α due to interband electronic transitions is proportional to $(\hbar\omega - \mathcal{E}_g)^{1/2}$, whereas for a GS this dependence is essentially different: $\alpha \sim \omega^{3/2}$.

Using subtle methods of phase transition theory, A.A. Abrikosov (1974, 1975) has investigated the influence of many-electron effects on the energy spectrum and charge carrier interaction and demonstrated that in the region of large wavevectors $k > a_{eB}^{-1}$ and energies $\mathcal{E} > \mathcal{E}_{eB}$(region I) perturbation theory is applicable for the parameter $(a_{eB}k)^{-1}$ and the formulas (2.22) and (2.23) are valid. In the region of wavevectors $10^{-2}a_{eB}^{-1} < k < a_{eB}^{-1}$ and energies $10^{-4}\mathcal{E}_{eB} < \mathcal{E} < \mathcal{E}_{eB}$ perturbation theory and the formulas (2.22) and (2.23) do not hold - this is the region of what is known as tight binding (region II). In this region the electron and hole dispersion relations are quadratic, but the effective masses change; the values of these masses should be determined experimentally. At very small wavevectors $(k < 10^{-2}a_{eB}^{-1})$ and energies $(\mathcal{E} < 10^{-5}\mathcal{E}_{eB})$ (region III) the dispersion relations of the one-electron approximation transform appreciably.

For the GS that are known, the typical value of the Bohr electron energy $\mathcal{E}_{eB}$ is $\simeq 1$ meV, so region III is practically inaccessible for experimental studies. In all of the diamond-like zero-gap semiconductors investigated to date, the impurity concentration is not lower than $\simeq 1\times10^{15}cm^{-3}$ and, correspondingly, the smallest Fermi energy is $\mathcal{E}_F > 1$ meV. All the physical characteristics of crystals are therefore determined by the electrons belonging to the energy region I $\mathcal{E} > \mathcal{E}_{eB}$ for which the k linear terms involved in the dispersion relation are very small. As far as the heavy holes in the region of wavevectors $a_{eB}^{-1} < k < a_{eB}^{-1} m_h/m_e$ and energies $\mathcal{E}_{eB} < \mathcal{E} < \mathcal{E}_{hB}$ are concerned, the k linear terms due to exchange interaction are essential and should be taken into account.

The exchange interaction affects also the energy spectrum of semiconductors with a narrow energy gap. Gelmont (1975) has considered the effect of the exchange interaction on the spectrum of narrow-gap semiconductors for the case of the two-band model (see Eq.(2.16)) in which the conduction band and the light-hole band are taken into account and the heavy-hole dispersion relation is quadratic and isotropic: $\mathcal{E}_h(k) = \hbar^2k^2/(2m_h)$. The most interesting result of the calculation is that owing to the exchange interaction the dispersion relation of heavy holes becomes nonquadratic at sufficiently large wavevectors satisfying the condition $kP \gtrsim \mathcal{E}_g$. In the region of small wavevectors the hole dispersion relation is quadratic:

$$\mathcal{E}_h(k) = \frac{\hbar^2k^2}{2}\left(\frac{1}{m_h} + \frac{2\sqrt{2}\,e^2P}{5\sqrt{3}\pi\hbar^2\kappa\mathcal{E}_g}\right), \qquad (2.27)$$

where m_h is the heavy-hole effective mass value obtained when account is taken of remote energy bands.

It follows from the expression (2.27) that for narrow energy gaps ($\mathcal{E}_g \lesssim 30$ meV) the effective mass of heavy holes near the top of the valence band is determined by the exchange interaction. Even with the gap $\mathcal{E}_g = 200$ meV, which corresponds to the solid solution $Hg_{1-x}Cd_xTe$ with x=0.26 (and, approximately, to InSb), the heavy-hole effective mass correction due to exchange interaction constitutes 15%.

A number of experimental investigations have been carried out to ascertain the role of many-electron effects in transport and optical phenomena. Specifically, attempts have been made to reveal the influence of the exchange interaction on the heavy-hole dispersion relation.

V.I. Ivanov-Omskii et al. (1983) have investigated, on p-HgTe samples, the reflection of light due to free charge carriers and found the effective mass of heavy holes to depend on their concentration p. Despite a considerable experimental error in the determination of the mass m_h, reaching 25%, it has been revealed that m_h tends to increase monotonically with p. This increase of m_h agrees qualitatively with the predictions of a theory which takes into account the exchange terms in the hole dispersion relation.

Analysis of interband-induced light absorption spectra in a magnetic field for HgTe and other materials permits no definite conclusion to be drawn as to the role of the exchange effects. It may currently be stated that so far no unambiguous and incontrovertible facts have been available which bear testimony to an appreciable contribution to the hole dispersion relation by the k linear terms due to exchange interaction. The energy range over which these terms should, according to theoretical estimates, be determinative is of the order of several milli-electron volts. Quite probably, the electron spectrum at these energies is highly distorted by the perturbations arising from the lattice defect potential. In consequence, the aforementioned energy interval proves particularly difficult to explore.

We have already mentioned above that the possibility of a spontaneous creation of electron-hole pairs in GS leads to an anomaly in the dielectric constant: For the region of energies near the band degeneracy point κ is very large (Liu and Brust, 1968 a, b). At temperature T=0 K and in the absence of donors, i.e., when there are no band electrons, this peculiarity of the dielectric constant leads to effective screening of impurity ion charges by the valence electrons which have an opportunity to make transitions between the valence band and the conduction band that touch; these transitions are said to be virtual[16)]. The absence of an energy gap in zero-gap semiconductors leads to creation of virtual excitons - a short-lived electron and a short-lived hole. The lifetime of an exciton is short, but excitons are generated continuously. Therefore, there is always a certain number of virtual electrons that screen charges. Real electrons, excited into the conduction band from donors or the valence band, say when the temperature is increased, suppress the singularity in the dielectric constant which is associated with the screening of charges by virtual electrons. This means that as the electron concentration of the conduction band is increased the dielectric anomaly should weaken, i.e., κ should decrease in magnitude.

The unusual behavior of the dielectric constant in gapless semiconductors, manifest in the enhancement of charge screening,should have some effect on the kinetic coefficients. Indeed, an enhancement of the impurity-center charge screening should lead to a decrease in the effective scattering cross-section, i.e., to a decrease of the probability that electrons will be scattered on an impurity center. In turn, this should result in an increase in electron mobility.In the purest HgTe crystals the electron mobility, calculated without allowing for the singularity of the dielectric constant for the case of electron scattering on impurity ions (which prevails at low temperatures $T \approx 4$ K), is almost an order of magnitude lower than the values found experimentally. Real HgTe crystals contain at least $\approx 10^{15} cm^{-3}$ donors. In electron mobility calculations one has therefore to allow for the weakening of the dielectric anomaly by conduction band electrons.

16) Virtual transitions are transitions of a system of particles from one state to another, which arise from the creation and annihilation of virtual particles. The latter exist only in intermediate, virtual states of short duration and cannot be recorded. Virtual transitions and states for a real system of particles are conventional and ambiguous (the process of transition of a system from the initial to the final state may be described as transitions via different virtual states). However, it is convenient to use these transitions in describing a variety of quantum processes, specifically particle interaction. Thus an interaction of two electrons is described as a process in which one electron emits and the other electron absorbs a virtual photon.

A calculation by Gelmont et al. (1971) of the mobility of electrons as a function of their concentration for HgTe at 4.2 K has shown that theory gives a good fit to the decrease in electron mobility with increasing electron concentration. For the purest samples the results of the calculation, in which interband screening is taken into account, practically tally with experimental data. It has turned out that interband screening plays a noticeable role up to electron concentrations $n \approx 10^{18} cm^{-3}$.

Of indubitable interest is the problem of how the contribution of the dielectric anomaly to electron mobility varies in going from a GS to a normal semiconductor. Such a transition occurs, for example, in solid solutions $Hg_{1-x}Cd_xTe$ when the Cd content is varied. At $x \simeq 0.16$ and at 4.2 K the light-hole band touches the degenerate bands at the point $\vec{k}=0$, and $\mathcal{E}_g=0$. Therefore the effective mass of electrons near the conduction band edge, m_e^*, (see Eq.(2.20)) should be close to zero at $x=0.16$ and, consequently, the electron mobility should be at a maximum. If one takes account of the interband screening, the theoretical $\mu_e(x)$ curves for a constant electron concentration give a good fit to experimental data. The role of interband screening manifests itself in the $\mu_e(x)$ curve being sharply assymmetric with respect to the transition point $x=0.16$: In gapless semiconductors, interband screening largely increases the electron mobility, whereas for a semiconductor with a gap at $x>0.16$ the contribution of interband screening is negligibly small.

2.3.7. Band Structure Readjustment under a Magnetic Field and Pressure

The equality to zero of the energy gap between the conduction and valence bands in GS generates a number of anomalies in transport phenomena. These peculiarities of the conductivity, Hall effect, magnetoresistance, etc. arise primarily because the electron and hole concentrations in zero-gap semiconductors may vary over a wide range under the influence of comparatively weak external factors such as electric and magnetic field, pressure, temperature, etc. This effect is two-fold: The electric fields, just as temperature, immediately affects the free charge carriers; the magnetic field and pressure alter the size of energy gaps, and this, in turn, leads also to alteration of the number of charge carriers.

In semiconductors with a wide gap between the conduction and valence bands, moderate electric fields, which do not suffice to excite electrons from the valence band into the conduction band, bring about a change in the energy distribution of electrons, raise their mean energy, and thus influence their mobility. As a result, the current versus voltage characteristic of the semiconductor deviates from the linear (ohmic) characteristic. In pure gapless semiconductors, interband transitions should begin in the weakest electric fields. Therefore the current versus voltage characteristic should not at all have any linear segment corresponding to Ohm's law. In addition to this, an increase in the mean energy of electrons in an external electric field leads to an increase in impact (collision) ionization rate[17)] and thus results in an increase in elec-

17) Impact ionization is the ionization of lattice atoms or impurities in the course of which valence band impurity level electrons are "dislodged" into the conduction band by free (band) electrons accelerated by electric field.

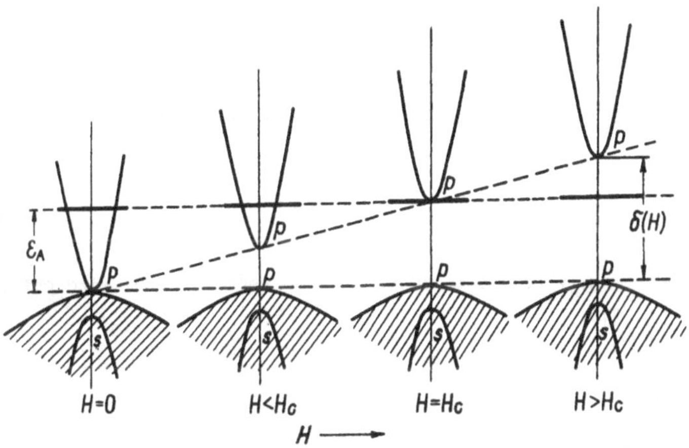

Fig.17. Variation of the band structure of a GS in magnetic field H. $\mathcal{E}_A$ is the energy of the acceptor level, δ(H) is the gap between the conduction and valence bands.

tron concentration. For this reason, the current-voltage characteristic should be grossly nonlinear. And this is what was observed in experiments on HgTe-type gapless semiconductors.

As is well known, the presence of a nonlinear current versus voltage characteristic is of great interest from a practical point of view. The nonlinear characteristic may be utilized, for example, to stabilize the amplitude of the output signal in devices employed to amplify and generate electromagnetic oscillations.

Unlike the electric field which immediately changes the number of charge carriers, the magnetic field causes the energy spectrum to readjust; namely, it opens a gap between the conduction and valence bands and, as a result, the electron and hole concentration changes too. The interaction of electrons with the magnetic field alters their energy and thereby leads, in particular, to displacement of the band edges. In normal semiconductors the gap between the valence and conduction bands unusually increases somewhat. In gapless semiconductors the p-symmetry bands in contact split up and thus a gap arises between them (see Fig.17).

Let us see how an electron moves and how its energy spectrum transforms in a magnetic field. Any charged particle, in particular an electron of charge e and velocity $\vec{v}$ in a homogeneous magnetic field $\vec{H}$ directed, say, along the z axis by the action of the force $e[\vec{v},\vec{H}]$, moves in a circle in the xy plane perpendicular to the magnetic field if the velocity component v_z along the field is equal to zero. The radius of the circle, r, is equal to $m_o v_\perp/eH = v_\perp/\omega_c$, with m_o being the mass of an electron, $v_\perp$ the velocity of travel of an electron in the xy plane, $\omega_c = eH/m_o$ the cyclotron frequency, i.e., the frequency of rotation of an electron around the magnetic field direction.

The condition for the applicability of classical mechanics reduces to a requirement that permits one to talk of a definite path of a particle: The wavelength λ of a particle should be considerably less than the length L specifying its path, i.e.,

$$\lambda \ll L . \tag{2.28}$$

In a magnetic field the characteristic length L is the radius of a circle, r, and the condition (2.28) takes the form

$$\lambda \ll r . \tag{2.29}$$

Since $\lambda = \hbar/m_o v$ and the energy $\mathcal{E} = m_o v^2/2$, it is convenient to represent the condiction (2.29) as

$$\hbar\omega_c \ll \bar{\mathcal{E}} , \tag{2.30}$$

where $\bar{\mathcal{E}}$ has the meaning of the characteristic energy of a particle. In the case of the Boltzmann statistics $\bar{\mathcal{E}} \approx k_B T$, and in the case of the Fermi-Dirac statistics $\bar{\mathcal{E}} \approx \mathcal{E}_F$.

If the condition (2.30) is violated, i.e., the magnetic field H is so strong that $\hbar\omega \gtrapprox k_B T$, electron orbit quantization effects become substantial. The energy of an electron turns out to be

$$\mathcal{E} = (N + \frac{1}{2}) \hbar\omega_c + \frac{\hbar^2 k_z^2}{2m_o} , \tag{2.31}$$

where the quantum number N is equal to 0, 1, 2, ... The first term here is the energy in the xy plane normal to the magnetic field, the second term is the energy of conti-

Fig.18. Landau subbands in magnetic field $H=H_z$ (a) and the density of states in a magnetic field as a function of energy (b); a) the dashed line shows the energy band in the absence of a magnetic field; b) the dotand-dash line shows the density of states at H=0 ($\rho_0 \sim \mathcal{E}^{1/2}$).

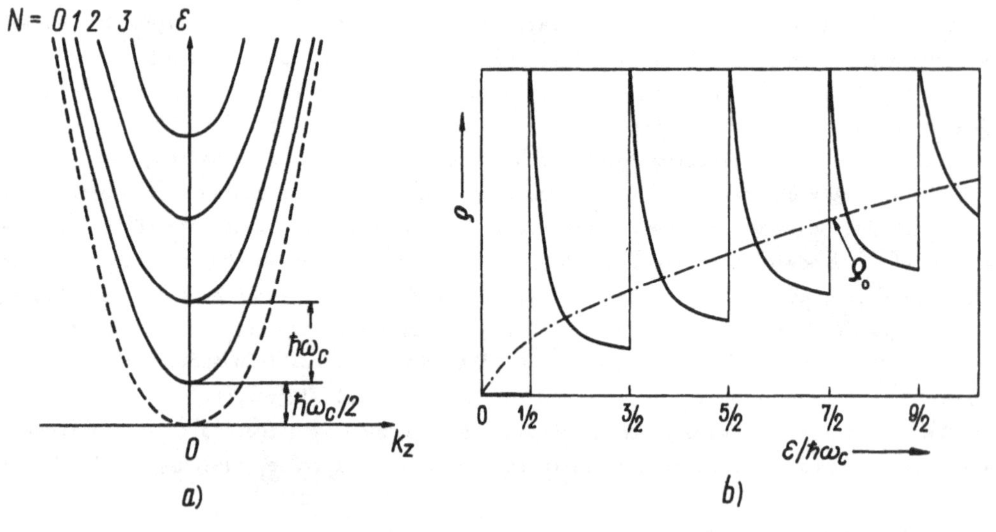

nuous movement along the z axis, i.e., along the magnetic field direction. Thus if parallel to the field $\vec{H}$ the electron energy spectrum remains continuous, whereas when perpendicular to $\vec{H}$ it is quantized, becomes discrete.

For an electron of effective mass m and cyclotron frequency $\omega_c = eH/m$ in a crystal, the three-dimensional energy band in $\vec{k}$ space with quasicontinuous distribution of levels split up by the magnetic field into one-dimensional magnetic subbands, or Landau levels (Fig.18a)), the distance between which is equal to $\hbar\omega_c$. Since the magnetic field sort of gathers the states distributed uniformly in the band into discrete Landau levels, the density of electronic states, ρ, changes appreciably: In a magnetic field the quantity ρ increases with H.

Figure 18b) presents the density of states, ρ, as a function of energy. Near the bottom of each Landau subband ($k_z=0$) ρ increases indefinitely. Since ρ increases with H, the Fermi energy $\mathcal{E}_F$ in the ultraquantum limit ($\hbar\omega_c >> \mathcal{E}_F$), when the electrons populate only the lower Landau subband, decreases with increase in H; that is, the Fermi level descends to the bottom of the conduction band.

The zero Landau level (N=0) is located $\hbar\omega_c/2$ higher than the band edge in the absence of a magnetic field (Fig.18a)), i.e., the magnetic field cause the band edge to shift upward in energy by $\hbar\omega_c/2$. The same thing happens to the valence band edge when the valence band is quantized; that is, the band is displaced downward.

As a result, the gap between the conduction and valence bands in a magnetic field increases in the case of parabolic bands by

$$\delta(H) = \frac{\hbar}{2}(\omega_{ce} + \omega_{ch}), \tag{2.32}$$

with ω_{ce} and ω_{ch} being the cyclotron frequencies of electrons and holes. This increase of the gap may be detected experimentally by studying the displacement of the fundamental light absorption edge due to interband transitions in a magnetic field.

Besides, a magnetic field causes the spin-degenerate levels to split up into two spin sublevels. For a free electron the energy difference of the split levels, $\Delta\mathcal{E}$, is equal to $\mathcal{E}_\uparrow - \mathcal{E}_\downarrow = \mu_B H$, with μ_B being a Bohr magneton. For crystals in which the spin-orbit splitting of bands, Δ, is large and the charge-carrier effective mass m is small, the spin splitting of levels is considerably larger:

$$\Delta\mathcal{E} = |g|\,\mu_B H \quad , \tag{2.33}$$

where the spectroscopic (Landé) splitting factor $|g|$ in semiconductors amounts to tens and even hundreds. Allowing for the spin splitting of levels, the energy of an electron in a quantizing magnetic field should therefore be written in the form

$$\mathcal{E} = \frac{\hbar\omega_{ce}}{2}\,(N + \tfrac{1}{2}) \pm \frac{g}{2}\mu_B H + \frac{\hbar\omega_{ce}}{2}\,(N + \frac{1}{2} \pm \frac{gm_e}{2m_o}) + \frac{\hbar^2 k_z^2}{2m_e} \quad . \tag{2.34}$$

In this case the gap between the bands which arises in a magnetic field is

$$\delta(H) = \frac{\hbar\omega_{ce}}{2}\,(1 - \frac{|g|m_e}{2m_o} + \frac{\hbar\omega_{ch}}{2}\,(1 - \frac{|g|m_h}{2m_o}) \quad . \tag{2.35}$$

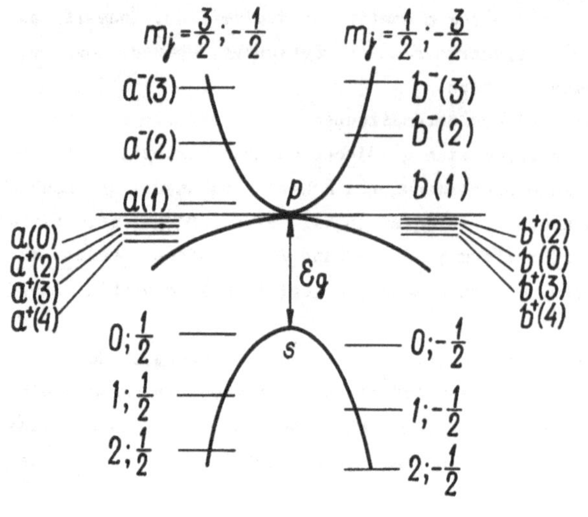

Fig.19. Diagram of the levels of a GS such as HgTe in a magnetic field

In GS with accidental degeneracy of the bands whose extrema lie at different values of the wavevector $\vec{k}$ (see Fig.10e)), the energy gap arising in a magnetic field is determined also by the expression (2.35). The situation is much more complicated in the case of zero-gap semiconductors such as α-Sn or HgTe, where the degeneracy of the conduction and valence bands is fundamental, i.e., is due to crystal symmetry. In the expression for the energy of degenerate-band charge carriers, terms appear which are dependent on the magnetic field and, in addition to the band parameters γ_1, γ_2, γ_3 (see Eq.(2.10)), a new parameter k appears which determines the g-factor of charge carriers.

With neglect of the nonsphericity of the isoenergetic surfaces, i.e., if one sets $\gamma_2=\gamma_3=\bar{\gamma}$ the energy levels in a magnetic field at $k_z=0$ group into two series, a and b, of nonequidistant levels according to the values of the projection m_j of the quantum number j of the total angular momentum. Series a involves levels with $m_j = 3/2, \ -1/2$; series b involves levels with $m_j = 1/2, -3/2$ (Fig.19). The energies of the levels for series a and b are determined by the expressions (Luttinger, 1956)

$$\mathcal{E}_{a,b}(N) = \hbar\omega_{co}\ \tilde{\mathcal{E}}_{a,b}(N) \tag{2.36}$$

where $\omega_{co} = eH/m_o$ is the cyclotron frequency of a free electron; $\tilde{\mathcal{E}}_{a,b}(N)$ is a function of the quantum number N, band parameters γ_1, $\bar{\gamma}$, and k.

For the quantum numbers N=0, N=1:

$$\tilde{\mathcal{E}}_a(0) = \frac{1}{2}(\gamma_1 - \bar{\gamma} - k),\ \tilde{\mathcal{E}}_a(1) = \frac{1}{2}(3\gamma_1 - 3\bar{\gamma} - k),$$

$$\tilde{\mathcal{E}}_b(0) = \frac{1}{2}(\gamma_1 + \bar{\gamma} - k),\ \tilde{\mathcal{E}}_b(1) = \frac{3}{2}(\gamma_1 + \bar{\gamma} - k). \tag{2.37}$$

Each of the series of levels, a and b, at N≥2 consists of two groups of levels $\mathcal{E}^+_{a,b}(N)$ and $\mathcal{E}^-_{a,b}(N)$; the expressions for these groups are not quoted here. In the two-band approximation (only the conduction p-band and the light-hole s-band are taken into account) $k = -\mathcal{E}_p/6\mathcal{E}_g$. In the same approximation the spectroscopic splitting factor for the lowest-lying conduction band Landau level is

$$g = 2(k - 3\bar{\gamma}). \tag{2.38}$$

The Landau levels $\mathcal{E}_{a,b}(1)$ and $\mathcal{E}^{-}_{a,b}(N)$ at $N \geq 2$ are conduction band levels, and the Landau levels $\mathcal{E}_{a,b}(0)$ and $\mathcal{E}^{+}_{a,b}(N)$ at $N \geq 2$ are valence band levels (Fig.19). For Landau levels with large numbers (N>>1), the spectroscopic splitting factor is

$$g = 4k - 4\bar{\gamma} - 2\gamma_1. \tag{2.39}$$

If one neglects the contribution of remote bands to the band parameters $\gamma_1, \bar{\gamma}$, k and allows only for the light-hole p-band closest to the degenerate s-bands, the following relations hold for these quantities:

$$\gamma_1 = 2\bar{\gamma} = 2k = -\frac{\mathcal{E}_p}{3\mathcal{E}_g}. \tag{2.40}$$

The expressions for the energies $\mathcal{E}_{a,b}(N)$ largely simplify in this approximation. The heavy-hole valence p-band turns out to be infinite-fold degenerate, i.e., all the Landau hole subbands merge into one plane band, a situation which corresponds to the infinite mass limit $m_h \to \infty$ (m_h assumes a finite value only if account is taken of the remote bands). The electron energy levels with $N \geq 1$ are described by the formulas

$$\mathcal{E}_a(N) = \hbar\omega_c^* \left(N - \frac{3}{4}\right),$$

$$\mathcal{E}_b(N) = \hbar\omega_c^* \left(N - \frac{1}{4}\right), \tag{2.41}$$

where $\omega_c^* = eH/m_e^*$, with $m_e^* = m_o/4\bar{\gamma}$ being the effective mass of an electron near the bottom of the conduction band.

In a sufficiently strong magnetic field a gap $\delta(H)$ arises between the conduction and valence band levels in zero-gap semiconductors. The lowest-lying conduction band level is $\mathcal{E}_a(1)$ (see Fig.19). According to the values of the parameters $\gamma_1, \bar{\gamma}$, and k, either the level $\mathcal{E}_a(0)$ or the level $\mathcal{E}_b^+(2)$ may turn out to be the topmost valence band level. For mercury telluride and its analogs, the energies $\mathcal{E}_a(0)$ and $\mathcal{E}_b^+(0)$ are rather close in magnitude. It may therefore be assumed that the displacement of the valence band top in a magnetic field is described by the movement of the level $\mathcal{E}_a(0)$. Then the gap $\delta(H)$ arising between valence and conduction bands in a magnetic field is determined by the distance between the levels $\mathcal{E}_a(1)$ and $\mathcal{E}_a(0)$ (i.e., by the energy level separation). According to the Equations (2.36) and (2.37),

$$\delta(H) = \hbar\omega_{co} (\gamma_1 - \bar{\gamma}). \tag{2.42}$$

In the two-band approximation (2.40) the gap that opens up in a magnetic field is

$$\delta(H) = \bar{\gamma}\hbar\omega_{co} = \frac{\hbar\omega_c^*}{4}, \tag{2.43}$$

i.e., is determined by the displacement of the lower electronic level only.

As analysis of experimental data shows, the smaller the gap $|\mathcal{E}_g|$ is, the better the Equation (2.40) works. This equation works fairly well for gaps $|\mathcal{E}_g| < 100$ meV which occur, for example, in $Hg_{1-x}Cd_xTe$ crystals with $0.1 \leq x \leq 0.2$. For HgTe the approximation (2.40) is very crude.

For small energy spacings $|\mathcal{E}_g|$ comparable with the gap $\delta(H)$ that arises, it is necessary to take into consideration the nonparabolicity of the conduction band. In the two-band approximation (allowing for the conduction p-band and the light-hole valence s-band) described in the absence of a magnetic field by the dispersion relation (2.16), the expressions for the energies of Landau electronic levels have the form

$$\mathcal{E}_a(N) = \frac{\mathcal{E}_g}{2} + \frac{|\mathcal{E}_g|}{2}\left[1 + \frac{4\hbar\omega_c^*(N - \frac{3}{4})}{|\mathcal{E}_g|}\right]^{1/2},$$

$$\mathcal{E}_b(N) = \frac{\mathcal{E}_g}{2} + \frac{|\mathcal{E}_g|}{2}\left[1 + \frac{4\hbar\omega_c^*(N - \frac{1}{4})}{|\mathcal{E}_g|}\right]^{1/2}. \tag{2.44}$$

From the Equation (2.44), it follows that in the case of appreciable conduction band nonparabolicity in the approximation (2.40), the gap determined by the shift of only the lower electronic level $\mathcal{E}_a(1)$ is not a linear (by contrast with the parabolic band case) function of the magnetic field:

$$\delta(H) = \frac{|\mathcal{E}_g|}{2}\left\{\left(1 + \frac{\hbar\omega_c^*}{|\mathcal{E}_g|}\right)^{1/2} - 1\right\}. \tag{2.45}$$

In the limit $\mathcal{E}_g \to 0$ we have from the Equation (2.45)

$$\delta(H) = \left(\frac{1}{3}\frac{eH}{\hbar}P^2\right)^{1/2} = \frac{1}{\sqrt{3}}\frac{P}{L}, \tag{2.46}$$

where $L = (\hbar/eH)^{1/2}$ is the magnetic length specifying the region of localization of the wavefunction of an electron in a magnetic field.

To experimentally determire the gap $\delta(H)$ one has apparently to measure the phenomenon sensitive to gap value. It is clear that with increasing gap at invariable temperature the electron concentration will decrease, since the electrons will be "frozen out" into the valence band more and more intensely, i.e., will recombine with holes. It is therefore of interest to measure effects that depend on intrinsic carrier concentration, for example the Hall effect and magnetoresistance.

As has already been mentioned, a magnetic field groups quasicontinuous energy band levels into discrete Landau sublevels (see Fig.18) and thus alters the electronic density of states which becomes a function of magnetic field H. Evidently, the statistical properties of charge carriers should depend on the relation between the quantities $\hbar\omega_c$ (Landau level separation) and k_BT (level smearing), and also on the position of the Fermi level $\mathcal{E}_F$.

Of practical interest for gapless semiconductors such as HgTe is the interval of magnetic fields and temperatures in which the electron spectrum is quantized ($\hbar\omega_{ce} >> k_BT$, only the lowest-lying Landau subband is filled with electrons) and the quantization of the hole spectrum may be neglected ($\hbar\omega_{ch} < k_BT$). These conditions are easy to satisfy for HgTe crystals (with magnetic fields of moderate intensity), since the electron effective mass $m_e << m_o$ and the hole effective mass $m_h \approx m_o$.

From the neutrality equation for an intrinsic GS (n=p) in a magnetic field, one can find the field H dependence of electron concentration n. In zero-gap semiconductors the magnetic field, as it opens a gap $\delta(H)$ between conduction and valence bands, causes the concentration of intrinsic charge carriers to decrease. If the gap that has opened up is sufficiently large, $\delta(H) > \mathcal{E}_F$, the electrons and holes obey, like in a semiconductor with a gap, the Boltzmann statistics. In this case the electron concentration should change in a magnetic field in the following fashion:

$$n(H) = A \left(\frac{\hbar\omega_{ce}}{k_BT}\right)^{1/2} \exp\left[-\frac{\delta(H)}{2k_BT}\right] \qquad (2.47)$$

with A being a constant independent of H.

The Equation (2.47) tells us that in a sufficiently large magnetic field, with the gap $\delta(H) > k_BT$, the concentration n should practically decrease exponentially with increasing field. An exact exponent is the function $n(H)/H^{1/2}$. If one plots $\ln(nH^{-1/2})$ vs. H, the value of the gap $\delta(H)$ may be determined from the slope of the curve. The same thing may be done, as follows from the Equation (2.47), if one makes a plot of $\ln(nH^{-1/2})$ as a function of 1/T.

Hall effect measurements on pure HgTe crystals (Giriat et al., 1975) have shown that at temperatures above 60 K, i.e., in the region of intrinsic conductivity, the Hall coefficient |R| is nearly constant in magnetic fields up to 150 kOe and increases with field at H>150 kOe (Fig.20). For HgTe, where the electron mobility is many times higher than the hole mobility, the Hall coefficient |R| in the intrinsic conductivity region is equal to 1/en; so an increase of

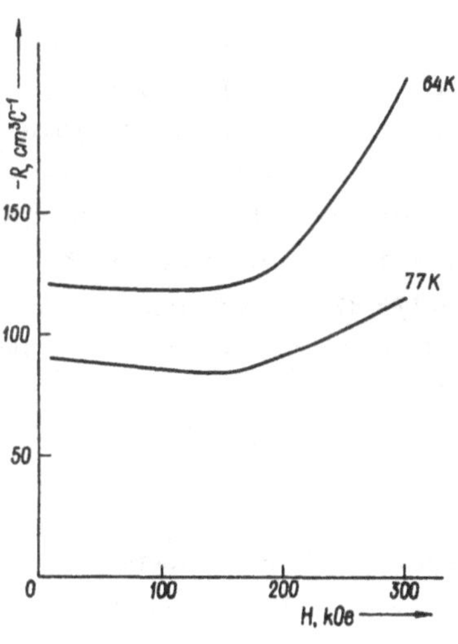

Fig.20. Magnetic field dependences of Hall coefficient for a HgTe crystal in the intrinsic conductivity region of temperatures 64 and 77 K

the Hall coefficient indicates that the electron concentration n diminishes with increasing H.

From the slope of the straight lines describing $\ln(|R|^{-1}H^{-1/2})$ vs. H, one obtains the value of the gap between the conduction and valence bands, $\delta(H) \simeq (5 \div 6)\hbar\omega_{co}$. Calculating $\delta(H)$ according to the approximate formula (2.43) leads to the value $\delta(H) = 10\hbar\omega_{co}$ (for $\mathcal{E}_p$ = 18 eV and $\mathcal{E}_g$ = -300 meV), which differs appreciably from the counterpart found from the Hall effect. But the more accurate formula (2.42) using γ_1, $\bar{\gamma}$ values determined from magnetooptical investigations (Gelmont et al., 1976, b) gives $\delta(H) \approx 6\hbar\omega_{co}$. Measurements of the magnetophon resonance in high electric and magnetic fields have yielded $\delta(H) = 5.4\ \hbar\omega_{co}$, in good agreement with Hall effect measurement data.

The occurrence of a gap in a magnetic field may be detected also from the variation of the magnetoresistance, which should increase abruptly with decrease in charge carrier concentration. Measurements of the longitudinal magnetoresistance ρ_{zz} (the magnetic field is parallel to the current and is directed along the z axis) have shown that in field H>150 kOe an increase in $\rho_{zz}(H)$ occurs which is close to exponential. Treatment of the $\rho_{zz}(H)$ curves with the help of the formula

$$\rho_{zz} \sim \exp \frac{\delta(H)}{2k_BT} \tag{2.48}$$

has yielded $\delta(H) = (7 \text{ to } 8)\hbar\omega$ at temperatures 64 to 77 K. This somewhat larger $\delta(H)$ value obtained from Hall effect measurements comes from the fact that the magnetoresistance, in contradistinction to the Hall effect, depends on the electron mobility which in a quantizing magnetic field varies with H. Therefore the expression for the magnetoresistance involves a factor that has a power-type field dependence: $(\hbar\omega_{ce}/k_BT)^r$, where the parameter r is determined by the scattering mechanism.

Thus the dependence of the magnetoresistance, $\rho_{zz}(H)$, should be sharper than that of the Hall coefficient, $R(H) \sim 1/n(H)$. The power-dependent preexponential factor in $\rho_{zz}(H)$ introduces a particularly appreciable error into the calculated gap value $\delta(H)$ if $\delta(H)$ only slightly exceeds k_BT. It is such a situation that occurred in measurements of $\rho_{zz}(H)$ at 64-77 K and in the field range 150-300 kOe (see Fig.20) (Giriat et al.,1975).

Like a magnetic field, hydrostatic pressure may turn a GS into a normal semiconductor. Qualitatively, this transformation is, however, of a different kind. Hydrostatic pressure does not break crystal symmetry and, therefore, does not cause the p-symmetry bands to split up - the degeneracy of these bands persists. Under hydrostatic pressure, in contrast to the effect of a magnetic field, the gap opens up not between two p symmetry bands but between p- and s-bands. As already stated in the foregoing, with increase in pressure $\mathcal{P}$ the light-hole band moves upward in energy, approaching the degenerate p-bands. In this case the gap between the s- and p-bands, $\mathcal{E}_g = \mathcal{E}(s) - \mathcal{E}(p)$ varies practically in a linear fashion with increasing pressure:

$$\mathcal{E}_g = \mathcal{E}_g(0) + \beta\mathcal{P} \quad , \tag{2.49}$$

where $\mathcal{E}_g(0)$ is the value of the gap at atmospheric pressure, and the pressure coefficient $\beta = d\mathcal{E}_g/d\mathcal{P}$ at k=0 is positive for all the semiconductors that are known. Thus

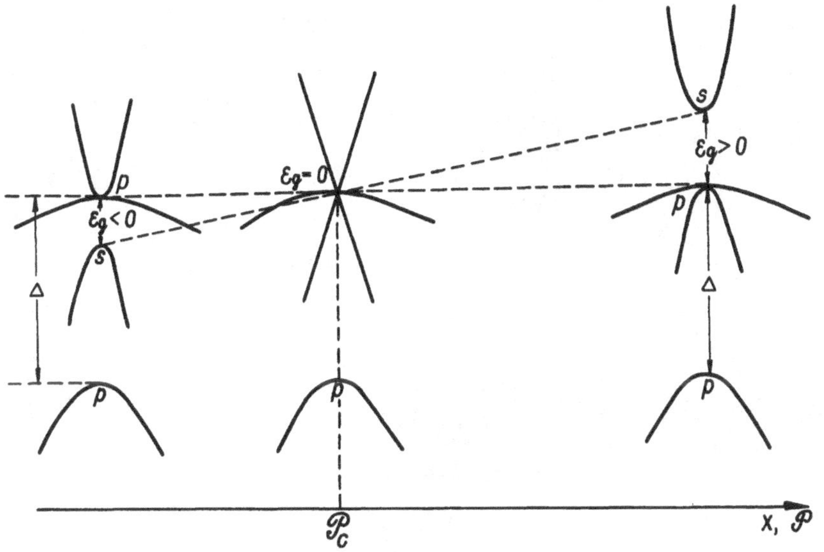

Fig.21.Reconstruction of the band structure of a GS under hydrostatic pressure $\mathcal{P}$ or when the composition of the solid solution (e.g., the Cd content (x) of $Hg_{1-x}Cd_xTe$) is varied

for GS with $\mathcal{E}_g(0)<0$ the quantity $|\mathcal{E}_g|$ decreases with increasing $\mathcal{P}$ and goes to zero at some pressure (the s- and p-bands touch (Fig.21)):

$$\mathcal{E}_g(\mathcal{P}_c) = 0. \tag{2.50}$$

When the pressure is increased further, i.e., $\mathcal{P}>\mathcal{P}_c$, an inversion of the bands occurs: The s-band, as it moves upward, becomes a conduction band, and the conduction p-band turns into a valence band. As this takes place, a gap $\mathcal{E}_g>0$ arises between the s- and p-bands; the value of this gap increases with pressure. Thus at pressure $\mathcal{P}=\mathcal{P}_c$ an electronic phase transition of a gapless semiconductor to a semiconductor with a gap occurs. For $\mathcal{P}=\mathcal{P}_c$ ($\mathcal{E}_g=0$) the dispersion relations of electrons and holes are linear (see Eq.(2.18)), the effective masses are minimal, and the mobilities maximal.

According to the experimental data of different researchers, the pressure coefficient β of the gap $\mathcal{E}_g$ in HgTe lowers somewhat as the temperature is decreased from (10 ± 5) meV/kbar at T=300 K to (8.5 ± 0.5) meV/kbar at T≈4 K. Hence it follows that at 4.2 K the gap $\mathcal{E}_g$ in HgTe should disappear at $\mathcal{P}_c$ = 35 kbar ($\mathcal{E}_g(0)$ = -0.3 eV). At much lower pressures $\mathcal{P}_{st}\simeq$ 13 kbar, however, HgTe turns out to undergo a structural phase transition due to which the zincblende lattice transforms into a hexagonal lattice such as that of cinnabar, HgS. For $Hg_{1-x}Cd_xTe$ crystals the value of the pressure $\mathcal{P}_{st}$ of structural phase transformation is somewhat larger than that for HgTe. Thus for $Hg_{1-x}Cd_xTe$ with x = 0.15 $\mathcal{P}_{st}\simeq$16 kbar. On the other hand, the pressure $\mathcal{P}_c$ of the GS-normal-semiconductor electronic phase transition decreases rather rapidly with increase in the Cd content of the solid solution. For crystals with x $\gtrsim$ 0.09, the electronic transition at helium temperatures sets in earlier than the structural transition does, i.e., $\mathcal{P}_c < \mathcal{P}_{st}$.

It is noteworthy that hydrostatic pressures bring about not one but at least two (thus far only two structural phase transitions have been detected) structural phase

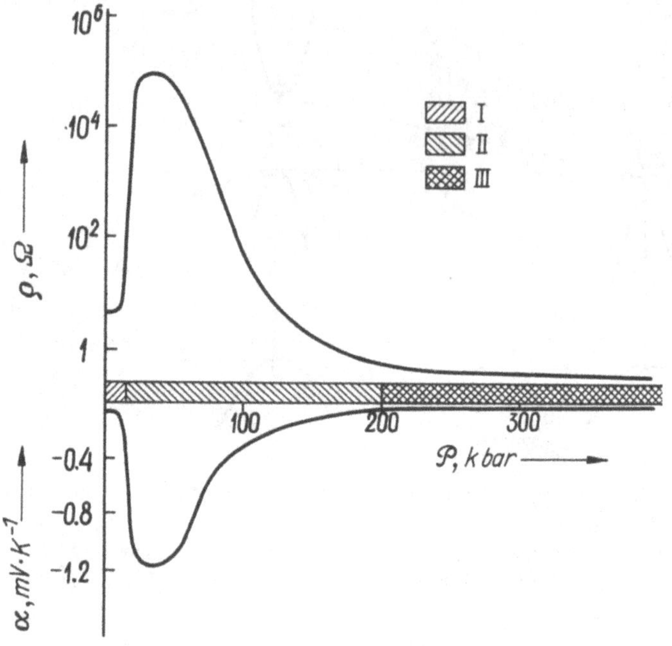

Fig.22. Dependence of revistivity ρ and thermal e.m.f. on hydrostatic pressure $\mathcal{P}$ for HgTe. I - III refer to the regions of a GS, a semiconductor, and a metal respectively.

transitions, accompanied by a readjustment of the electron spectrum, in gapless semiconductors. Figure 22 presents the pressure dependences of electrical resistivity ρ and thermal e.m.f. α for zero-gap semiconductors HgSe. At pressure $\mathcal{P}_{1st} \approx 10$ kbar the cubic HgSe turns (in the various samples the resistivity increases by 4-6 orders of magnitude) into a hexagonal semiconductor with a gap $\mathcal{E}_g \approx 0.7$ eV at 300 K. The gap then decreases in value and vanishes (the resistivity falls abruptly) at pressure $\mathcal{P}_{2st} \approx 200$ kbar - a metal forms which exhibits normal resistivity and thermal e.m.f. values (Tsidilkovski, 1982, 1983).

CHAPTER 3

IMPURITIES IN GAPLESS SEMICONDUCTORS

3.1. The Problem of Residual Electron Concentration

It is most advantageous to study features peculiar to the electron spectrum of GS at low temperatures, when the mean thermal energy of charge carriers does not exceed the characteristic electronic quantities - the Bohr energy, the Fermi energy, and the energy spacings between levels. According to the formulas (2.7) and (2.8), when the temperature is lowered down to $T \to 0$ the intrinsic charge carrier concentration n_i, in the vicinity of the degeneracy point of the bands in a zero-gap semiconductor should decrease in keeping with the $n_i \sim T^{3/2}$ law. However, numerous Hall effect investigations on the purest HgTe and α-Sn crystals have shown that for $T \lessapprox 10$ K appreciable deviations from this law occur, the electron concentration at $T \to 0$ tending to a constant value of n_o. The limiting residual concentration n_o for different crystals is approximately $n_o \lessapprox 1 \times 10^{15} cm^{-3}$. It may evidently be assumed that one of the possible causes for the existence of the residual electron concentration n_o is impurities. In this context we wish to discuss briefly the impurity-induced changes in the electron spectrum of an ideal crystal.

At low temperatures the electronic properties of semiconductors are determined primarily by electrically active impurities, viz., donors and acceptors. The atom of a donor impurity in a crystal is comparatively easy to ionize and becomes positively charges, and the electron is excited into the conduction band and can participate in charged,transport. The acceptor impurity is capable of capturing a lattice atom electron and acquires a negative charge as it does so, a hole arising in the valence band. When the temperature is sufficiently lowered the electron returns to the donor, and the hole returns to the acceptor. The electrons are said to be frozen out into donors and holes. On the energy scheme of the semiconductor, the donor and acceptor levels are located in the forbidden band (Fig. 10b)). The discreteness of these levels means that the electron is localized at the donor (the hole at the acceptor) and cannot move through the crystal.

As the concentration of impurity centers, say donors, is increased, the electron wavefunctions of adjacent donors start overlapping and a region of allowed states of finite energy width - an impurity band - arises. Attention should be immediately paid to the fact that the impurity band differs radically from the allowed energy bands of the crystal. What counts is not that, owing to the lower impurity atom content compared to the number of lattice atoms, the impurity band is narrower than the fundamental bands of the crystal. The chief distinction lies in the impurity atoms being arranged randomly in the lattice.

The absence of periodicity in their arrangement does not, strictly speaking, allow one to exploit such important concepts of band theory as the dispersion relation, effective mass, etc., the introduction of which is possible only for a periodic potential. However, a number of fundamental concepts of band theory, when generalized in some way, hold good also for a disordered system, of which the impurity band is an example. Here belongs primarily the density of electronic states, $\rho(\mathcal{E})$. This quantity and the observables for a disordered system (conductivity, light reflection coefficient etc.) are related in the same way as the counterpart for ordered systems.

Using the concept of the density of states, it is possible to introduce, for a disordered system, the concept of allowed and forbidden bands: An allowed band is an energy interval in which the density of states, $\rho(\mathcal{E})$, is nonzero and continuous, whereas $\rho(\mathcal{E}) = 0$ in a forbidden band. A number of important features of the band scheme persist also for disordered systems such as an impurity band, but there are substantial differences as well. Thus, because of the random distribution of impurities, allowed bands do not have well-defined edges, in contrast to an ideal crystal. In these bands, "tails" arise which extend into the forbidden band. This occurs in semiconductors that contain a great deal of impurities, i.e., in heavily doped semiconductors, where the impurity band is so wide that it overlaps with the "host" band of the crystal.

The reason why the density of states in a heavily doped semiconductor does not vanish at the energy band edge, i.e., the well-defined band edges disappear and the region of allowed energies shifts into the interior of the forbidden band, is as fol-fows. Even when impurities are distributed uniformly, on an average, in a sample there are always local fluctuations of the number of impurity atoms in some separated volume. This leads to a local decrease or increase in electron energy compared to its mean value. In those regions where the energy of electrons lowers, a density of states tail arises. Thus the displacement of the conduction (valence) band edge in heavily doped semiconductors is due to the interaction of electrons (holes) with impurity atoms that are distributed nonuniformly in the bulk of the crystal.

We now return to the problem concerning the causes responsible for the existence of the residual electron concentration n_o when the temperature of GS is lowered. Originally it was assumed that the residual concentration in HgTe arises from a small overlap of the conduction and valence bands (see Figs. 13 and 23b))

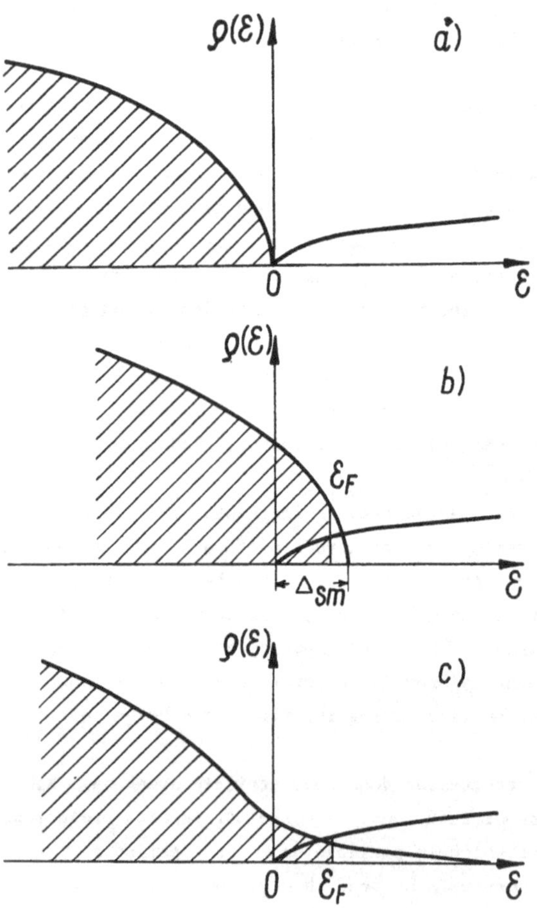

Fig.23. Density of states for an ideal GS (a), for a semimetal (b), and for a GS with $m_e \ll m_h$, in which the band overlap is due to a fluctuating potential (c). The filled states are shaded.

(Galazka, 1970; Sologub et al., 1971). The material thus is essentially a semimetal and the electron concentration remains finite for $T \to 0$. The occurrence of a weak band overlap in crystals with zincblende structure that possesses no center of symmetry is possible because the dispersion relation involves small linear terms in the wavevector. These terms lead to the valence band maximum shifting somewhat from the point $\vec{k}=0$ in the <111> direction and upward in energy. For InSb, for example, this energy shift is very small, approximately 0.005 meV. No reliable experimental data is available for HgTe, a theoretical estimate of the semimetallic overlap value giving $\Delta_{sm} \lessapprox 0.3$ meV (Overhof, 1971). According to other estimates, the overlap Δ_{sm} due to valence band shift in zero-gap semiconductors with zincblende structure should range between 10^{-1} and 10^{-3} meV (Groves, 1971).

Originally, the absence of a freeze-out of electrons into donors in HgTe when the temperature is lowered down to 0.1 K was held to be a weighty argument in favor of the semimetallic band overlap and the intrinsic nature of the residual electron concentration (Sologub et al., 1971). However, it was shown later that a positively charged donor in GS with widely differing electron and hole masses $m_e \ll m_h$ is incapable of retaining an electron in the Coulomb potential well produced by it. So all the donors turn out to be ionized already at $T = 0$ K. Thus the absence of a freeze-out of electrons into donors in HgTe cannot supply proof of the band overlap being semimetallic.

By the value of residual concentration n_o it is possible to estimate Δ_{sm} for a semimetal. At $T = 0$ K the Fermi energy of a semimetal is $\mathcal{E}_F = (m_h/m_h+m_e)\Delta_{sm}$. On the other hand, for a degenerate electron gas, $\mathcal{E}_F = (\hbar^2/2m_e)(3\pi^2 n_i)^{2/3}$. Assuming the experimental value of residual electron concentration in HgTe $n_o = 7 \times 10^{14}$ cm^{-3} to be the intrinsic concentration n of a semimetal, one readily obtains $\Delta_{sm} \simeq 1$ meV. This value of Δ_{sm} exceeds the theoretical estimates cited above, and it may therefore be thought that the residual electron concentration in HgTe is assured not by the conduction and valence band overlap due to the linear terms involved in the hole dispersion relation.

This conclusion is corroborated by experimental data for α-Sn. It was found that in the purest α-Sn samples the minimum electron concentration $n_o = 4 \times 10^{14}$ cm^{-3} is due to donor ionization. The ionized donor concentration, equal to the electron concentration n_o, turned out to vary under uniaxial pressure (Roman and Ewald, 1968). Thus the concentration n_o is not intrinsic and, accordingly, is not due to semimetallic band overlap. In addition, it must be noted that for α-Sn whose lattice possesses a center of symmetry, in contradistinction to HgTe, linear terms in the hole dispersion relation are forbidden by symmetry conditions and the occurrence of a semimetallic overlap is not possible at all.

In 1983 Efros et al. (Ablyazov et al., 1983) proposed a new gapless-semiconductor model in which the residual electron concentration is also explained by an overlap of the conduction and valence bands. However, in its physical nature, this overlap has nothing in common with the semimetallic overlap discussed above. According to the model proposed, electrically inactive defects, i.e., those defects which do not serve as sources of electrons and holes (fluctuations in the composition of solid solutions such as HgCdTe, nonregular field of inner stresses in the crystal) produce a potential $V(\vec{r})$ which depends randomly on coordinates. The interaction of electrons with this potential distorts the shape of the conduction and valence bands at their edges (similar to the situation that occurs in heavily doped semiconductors), as a result of which an overlap of these bands arises (Fig. 23c)).

By the action of the fluctuating potential $V(\vec{r})$, the electronic system becomes spatially inhomogeneous; near the band edges, fluctuational potential wells arise in which electrons and holes may be localized. The point at which the conduction band and the valence band touch moves upward or downward in energy relative to its level in an ideal crystal. Energy levels for electrons and holes appear in the fluctuational wells that have formed. Because of the large difference between electron mass and hole mass ($m_e \ll m_h$), the width of the hole levels is sufficiently small, whereas the electron levels are highly smeared.

The density of states for the quadratic dispersion relation is $\rho(\mathcal{E}) \sim m^{3/2}\mathcal{E}^{1/2}$. Figure 23a) presents the function $\rho(\mathcal{E})$ for an ideal GS. The density of states of electronic levels in fluctuational wells, even if the smearing of these levels is not taken into account, is much less than the density of states of hole levels in valence band wells. The ration of these densities of states of the levels is proportional to $(m_e/m_h)^{3/2}$. For $m_e \ll m_h$ one may therefore take into account only the hole levels in fluctuational wells. They encompass the region of energies above the degeneracy point of the bands of an ideal crystal and emerge on the conduction band spectrum (Fig.23c)).

The density of states of hole levels decreases with energy into the interior of the conduction band according to the exponential law. In its nature, this density of states tail is similar to that arising in heavily doped finite-gap semiconductors from the non-uniform spatial distribution of impurity ions. However, it is in GS that the formation of a fluctuational density of states tail leads to effective overlapping of the valence band with the conduction band, in contrast to the very problematic band overlap which may be a consequence of the band structure of zincblende-type crystals without a center of symmetry. Essentially, the fluctuational model leads to the conclusion that the gapless state is peculiarly instable (in the statistical sense) with respect to the perturbing action of a random potential of any nature, and the density of states of a zero-gap semiconductor should be similar to the counterpart of a semimetal (cf. Figs. 23b) and 23c)).

Since, doubtless, real crystals always contain some electrically inactive defects or other which produce a random potential $V(\vec{r})$, it may be assumed that the described model of the band overlap due to the formation of fluctuational density of states tails should be inherent in any gapless semiconductor.

From a physical point of view, the fluctuational conduction and valence band overlap model appears to be very attractive. However, it would be able to describe the properties of only GS that contain no electrically active impurities, i.e., no donors and no acceptors. At the same time, as experiment shows, all of the gapless semiconductors investigated to date contain rather considerable amounts of donors and acceptors and are thus incapable of possessing intrinsic conductivity at low temperatures. Therefore the residual electron concentration n_o detected in α-Sn and HgTe is most likely to be due to the presence of donor- and acceptor-type impurities in all the crystals that have been studied. At $m_e \ll m_h$ the donors in GS are incapable of retaining electrons in their Coulomb well even at T = 0 K, so at the lowest temperatures there should be some amount of electrons in the conduction band. It is, however, deemed that a more important cause for the presence of residual concentration n_o is the band overlap due to fluctuations of the potential of charged donors and acceptors.

3.2. Impurities and Native Defects in Mercury Chalcogenides

In monatomic gray tin the interatomic bond is covalent, and in mercury chalcogenides it is of a mixed ionic-covalent nature. Proceeding from considerations based on the character of bonding and crystal structure, it is, generally speaking, possible to predict the behavior of some elements whose atoms form substitutional impurities in the lattice of II-VI crystals. Thus it ought to be expected that when replacing divalent Hg, Cd or Zn cations the Periodic Table group I elements should behave as acceptors (see sect. 1.5.). The group V elements, on replacing a Te or Se anion should act also as acceptors. If the cation is replaced by a group III element, and the anion by a group VII element, then the substitutional atoms should behave as donors.

Experiment has shown that these general considerations are borne out only in part. The I and III elements replacing metallic ions have, indeed, turned out to be electrically active, i.e., serve as acceptors or donors. But impurities of the elements belonging to groups V and VII, as a rule, do not affect the electrical properties of mercury chalcogenides. In HgTe and HgCdTe, dopants such as Cu, Ag, and Au are most frequently used. This allows samples with maximum hole concentrations $p \approx 10^{18} - 10^{19}$ cm^{-3} to be synthesized. Crystals with electron concentrations of the same order of magnitude are produced by doping with the group III elements In, Ga, Al (Harman, 1967; Long and Schmit, 1970; Harman and Melngailis, 1974).

Insofar as crystals of zero-gap and narrow-gap II-VI semiconductors with charge carrier concentrations $\lesssim 10^{17}$cm^{-3} go, the remain role in the formation of their electrical properties is played by native defects rather than foreign impurities. The point is that in contradistinction to the III-V compounds (InSb, GaAs, etc.), which possess practically ideal stoichiometry, departures from stoichiometry in crystals based on mercury chalcogenides are very appreciable. The shortage of mercury atoms, which is liable to be a consequence of cationic sublattice vacancies or interstitial chalcogen (Te, Se) atoms, leads to formation of acceptor-type defects. Interstitial mercury atoms or anionic sublattice vacancies should be donors.

Lattice defects that arise due to departure from stoichiometry manifest themselves in an entirely different way in HgTe and HgSe. As-grown (but not specially annealed) crystals or epitaxial layers of HgTe, HgCdTe, HgZnTe, and HgMnTe always possess hole conductivity (p-type crystals) with hole concentrations $p \simeq 10^{17} - 10^{18}$ cm^{-3}. But HgSe, HgCdSe, HgZnSe, and HgMnSe crystals are always gapless or narrow-gap n-type semiconductors, and it is noteworthy that it has not been possible to bring the electron concentration below a level of $n \approx 10^{16} - 10^{17}$ cm^{-3} whatever the annealing conditions. Extrinsic (n-type) crystals with these electron concentrations and an effective mass $m_e \approx 10^{-2} m_0$ are heavily doped semiconductors; that is, impurity states overlap substantially with the conduction band. Peculiarities of impurity states therefore do not show up in n-type crystals.

Considerable progress has been made in the investigations of HgTe. This success owes much to the discovery that annealing samples under the pressure of mercury vapor substantially reduces native defect concentration and thereby improves their electrical properties (Tsidilkovski, 1955; Harman, 1958; Rodot and Triboulet, 1962). At present, heat treatment of HgCdTe crystals at a certain temperature and at a certain pressure of the vapor of the constituents has been extensively used to approximate the material to stoichiometric composition and to govern the defect concentration.

As a result of numerous investigations, it has been found that the concentration of impurity-generated charge carriers in HgCdTe samples is at a minimum if the annealing temperatures lie within the limits $200 \le T_{ann} \le 350°C$. At $T_{ann} > 400°C$ the samples always turn out to be p-type crystals, and the concentration of acceptors produced by native lattice defects varies as a function of annealing temperature (the higher the temperature, the larger the acceptor concentration). A maximal equilibrium concentration of native acceptor-type defects in $Hg_{1-x}Cd_xTe$ is attained at annealing temperatures close to the melting point T_m of the solid solution. With increasing Cd content this concentration p^{max} gradually decreases: for HgTe (T_m = 670°C) $p^{max} \simeq 1.5 \times 10^{19} cm^{-3}$, for an alloy with x = 0.2 ($T_m \simeq 705°C$) $p^{max} \simeq 3 \times 10^{18} cm^{-3}$, for an alloy with x = 0.4 ($T_m \simeq 745°C$) $p^{max} \simeq 6 \times 10^{17} cm^{-3}$.

It is noteworthy that the range of possible continuous variation for the concentration of equilibrium defects - acceptors, or the region of homogeneity of $Hg_{1-x}Cd_xTe$ solid solutions (the region in which a stable one-phase solid solution exists and no second phase of some constituent precipitates) varies in a nonmonotonic fashion with increasing Cd content. Thus for HgTe samples annealed at 550°C the electron concentration increases from $4.7 \times 10^{18} cm^{-3}$ when the sample is saturated with mercury to $9 \times 10^{18} cm^{-3}$ when the sample is saturated with tellurium,i.e., by approximately a factor of 2. For $Hg_{1-x}Cd_xTe$ samples with x = 0.2 the region of homogeneity at the same annealing temperature 550°C ranges between 2×10^{17} and $3 \times 10^{18} cm^{-3}$, i.e., the concentration of equilibrium defects varies as a function of mercury vapor pressure by more than an order of magnitude. For solid solutions with x = 0.4 the region of homogeneity narrows again: At T_{ann} = 550°C the acceptor concentration varies from 1×10^{17} to $4 \times 10^{17} cm^{-3}$ (Strauss and Brebrick, 1970; Vydyanath, 1981; Vydyanath et al., 1981).

The annealing time during which a thermodynamic equilibrium is established between the sample and the vapor phase depends on annealing temperature and the prehistory of the sample (method of synthesis, initial defect concentration, impurity content).

Thus for $Hg_{1-x}Cd_xTe$ samples of thickness $d \le 0.4$ mm the annealing time at $T_{ann} \ge 500°C$ ranges between 16 and 24 hours, whereas at $T_{ann} \le 460°C$ this time is equal to two or three weeks. To best preserve the high-temperature equilibrium samples subjected to annealing are rapidly quenched, for example in ice water.

The problem concerning the role of native donor-type defects in HgTe has been debated for a rather long time.

It has been found that annealing specially undoped ingots with p-type conductivity at T_{ann} < 350 to 400°C under Hg saturated conditions permits production of n-type samples with electron concentration $n \approx 10^{14}-10^{15} cm^{-3}$. Proceeding from this fact, a number of authors of early papers (see Long and Schmit, 1970; Dornhaus and Numtz, 1976) held that, depending on heat treatment conditions, native lattice defects in annealed HgCdTe crystals may be of both the acceptor and donor type and the conversion of p-type samples to the n-type at T_{ann} < 350-400°C is due to native donor-type defects.

More conclusive is the point of view of many other researchers (Reynold et al., 1971 ; Schmit and Stelzer, 1978; Vydyanath, 1981) who hold that native lattice defects in annealed HgCdTe crystals may be acceptors only (they are most likely to be produced by mercury vacancies), and donors are foreign impurity atoms. It has been found that the electron concentration n in $Hg_{0.8}Cd_{0.2}Te$ crystals annealed at $T_{ann} < 360°C$ under Hg saturated conditions is independent of annealing temperature. The T_{ann} independence of concentration means that donors which furnish electrons are not native lattice defects in annealed samples and the electronic conductivity is due to foreign donor impurities.

This is borne out also by the following experimental fact. Some p-type $Hg_{0.8}Cd_{0.2}Te$ samples with hole concentrations $p \approx 10^{15}-10^{16} cm^{-3}$ did not turn to n-type crystals even upon annealing at temperatures below 350°C under Hg saturated conditions. If conversion of hole conductivity to electronic conductivity occurred owing to native donor defects it would be hard to explain why a number of samples cannot be converted to n-type crystals. It may be believed that the residual concentration of foreign acceptors in these samples exceeds that of donor defects (foreign donors), and n-type to p-type conversion is impossible, whatever the annealing temperature may be.

The following observation is of interest. Extrinsic $Hg_{0.6}Cd_{0.4}Te$ samples which were not doped specially transformed to n-type crystals at $T_{ann} < 350°C$ under Hg saturated conditions, but the conversion temperature turned out to be different for various samples. It is quite reasonable to assume that the difference in conversion temperatures is due to the presence of foreign rather than native donor impurities. It stands to reason that this conclusion applies to annealed samples; as-grown unannealed samples have not only acceptor-type but also donor-type native defects. This is confirmed by the experimental result that subsequent to annealing there is a decrease in both acceptor concentration N_A and donor concentration N_D. The donor concentration decreases by a much larger amount, so that the degree of compensation of the sample, N_D/N_A, may be lowered substantially.

We conclude this section by noting one more experimental fact which supports the conclusion that donors in annealed samples are produced by foreign impurities rather than native defects. Vydyanath et al., (1981) have studied the effect of different annealing conditions on hole concentration and mobility at 77 K in $Hg_{0.8}Cd_{0.2}Te$ and $Hg_{0.6}Cd_{0.4}Te$ samples. The annealing temperature and the pressure of Hg vapor above the samples was so chosen that the hole concentration at 77 K, $p = N_A - N_D$, was the same in all the samples. It turned out that the mobility of holes at 77 K in the samples was practically the same. Since what dominates at 77 K is hole scattering on ionized impurity centers, the mobility of holes depends on $N_A + N_D$. If not only the acceptor concentration N_A but also the donor concentration N_D varied as a function of annealing conditions, the hole mobility at the same concentration $N_A - N_D$ would certainly be different in distinct samples. The fact that the hole mobility is practically constant under different annealing conditions assuring the same concentration indicates that the native donor defect concentration of annealed crystals is negligibly small (Vydyanath et al., 1981).

Thus numerous investigations have shown that as-grown HgTe and HgCdTe crystals contain, as a rule,a considerable amount of native defects, both acceptors and donors, i.e., are compensated with hole concentrations $N_A - N_D \gtrsim 10^{17} cm^{-3}$. Annealing permits decreasing both the defect concentration and the degree of compensation N_D/N_A

3.3. Energy of Impurity States

In a finite-gap semiconductor the energy levels of an electron or hole captured by impurity atoms are discrete if they are located in the energy gap (Fig. 10b)) and correspond to localization of a charge carrier on an impurity atom. In GS where the gap between the conduction band and the valence band is absent, the impurity levels, i.e., the possible energy values of an electron or hole on an impurity center, land inevitably in the region of the continuous spectrum of allowed energy bands of an ideal crystal: The acceptor levels emerge in the conduction band, and the donor levels in the valence band (Fig. 10d)). These states may not be discrete, i.e., the electron and holes cannot produce bound donor and acceptor states on impurity atoms. This is so because, say, an electron localized on an acceptor acquires an opportunity to pass to a conduction band state of the same energy. The residence time of an electron on an acceptor, or the lifetime τ, here is already not infinite, in contrast to the case of an acceptor level located in the energy gap of a normal semiconductor. In consequence, according to the Heisenberg indeterminacy principle, the level should broaden (become smeared), i.e., acquire a finite width $\Gamma = \hbar/\tau$. This is a so-called natural level broadening[18].

Thus, since the impurity levels of GS are located on the background of the continuous spectrum of a perfect crystal, or, as is customary to say, are resonant, the states of an electron (hole) on an impurity are not completely bound but quasilocalized. Another feature of the impurity states in GS is an unusual inverted arrangement of donor and acceptor levels. Donor levels split off from the conduction band by the perturbing potential of a defect land in the valence band and settle down below acceptor levels which emerge in the conduction band. In semiconductors with a finite gap $\varepsilon_g > 0$, such a situation is liable to arise only for deep impurity levels (which in recombination processes play the role of nonequilibrium charge carriers) of an energy of the order of ε_g. Shallow levels which are responsible for the electron and hole distribution in energy and govern the transport characteristics of the semiconductors, arrange themselves in the habitual order: Donor levels lie beneath the bottom of the conduction band above the acceptor levels split off from the valence band.

The notion of a "shallow impurity level" as applied to gapless semiconductors has a different significance than for a usual semiconductor. The "shallow level" criterion for a semiconductor with a finite gap is of the form $\varepsilon_i \ll \varepsilon_g$, with ε_i being the energy of the bound state of an electron or of a hole on an impurity center. In GS the gap between the conduction and valence bands, ε_g, is equal to zero, and so this criterion becomes senseless. In the case of zero-gap semiconductors it is reasonable to consider as shallow those impurity levels which lie near the Fermi level and largely affect the charge carrier distribution in energy and transport phenomena. The presence of impurity levels on the background of the continuous spectrum leads to a number of peculiarities in the macroscopic electronic properties of GS, as will be discussed later in the text. And now we turn to the problem concerning the energy and spatial behavior of the wavefunction of an electron (hole) on an impurity center in GS.

The problem of finding the energy and wavefunction of an electron localized on an impurity can, in principle, be solved if the Schrödinger equation for an electron in an

18) Apart from this, when the number of acceptors is sufficiently large the wavefunction of holes on adjacent acceptors (or, for brevity, acceptor wavefunctions) overlap and the level diffuses into the acceptor band. In this case one talks of a concentrational broadening of the level.

ideal crystal involves the electron-impurity interaction potential $V(\vec{r})$ (see Eq.(1.23)) as a summand. In the absence of free charge carriers, the potential $V(\vec{r})$ for a charged donor or acceptor in a crystal with a dielectric constant κ is a Coulomb potential, $V(\vec{r}) = -e^2/\kappa r$. If there are free charge carriers which screen the charge of an impurity center, then $V(\vec{r}) = -(e^2/\kappa r)\exp(-r/r_o)$, with r_o being the screening radius. The Schrödinger equation for a four-fold degenerate p-symmetry band is a system of four coupled second-order equations in partial derivatives and is difficult to solve. With neglect of the non-sphericity of isoenergetic surfaces, the problem simplifies: It reduces to two uncoupled subsystem, each containing two ordinary differential equations.

Gelmont and Dyskonov (1972) have calculated the energy of the acceptor ground state for HgTe-type gapless semiconductors. It has been possible to obtain rigorous quantitative results only for a small ratio of electron and heavy-hole effective masses ($m_e \ll m_h$). In the limit $m_e \to 0$ and for a finite hole mass the acceptor level energy is

$$\mathcal{E}_A = \frac{2}{9}\,\frac{m_h e^4}{\kappa^2\hbar^2} \equiv \frac{4}{9}\,\mathcal{E}_{hB}\,, \tag{3.1}$$

with $\mathcal{E}_{hB}$ being the Bohr energy of a heavy hole. For the case of HgTe with $\kappa = 20$ and $m_h = 0.4\,m_o$ the Equation (3.1) yields $\mathcal{E}_A = 6$ meV.

Thus, when $m_e \to 0$, discrete acceptor levels appear on the background of the continuous conduction band spectrum. The wavefunction of the quasilocalized ground state of a hole and an acceptor at distance $r \approx a$ from the acceptor (a_{hB} is the Bohr radius of a hole) decreases exponentially with r. When far from the acceptor, with $r > a_{hB}$, the wavefunction decreases obeying the power law as $1/r^3$. But if one allows for the fact that the electron mass m_e is finite, then at large distances separating the acceptor, $r > \sqrt{a_{hB}a_{eB}}$ where $V(\vec{r}) \to 0$, the wavefunction does not decay but oscillates, just as it does in the free electron case, with a period of the order of $\hbar/\sqrt{m_e\mathcal{E}_A} \approx \sqrt{a_{hB}a_{eB}}$. The amplitudes of these oscillations are directly proportional to the electron mass to hole mass ratio.

The presence of wavefunction oscillations means that the acceptor level is notwell-defined but smeared. The level broadening Γ may be found by calculating the probability of a transition $W \sim 1/\tau$ from the acceptor state to the state of the continuous conduction band spectrum. The level width $\Gamma = \hbar W$ turns out to be proportional to the conduction band density of states. Calculations allowing for the nonlocality of the exchange potential yield

$$\frac{\Gamma}{\mathcal{E}_A} = 8\left(\frac{m_e}{m_h}\right)^{3/2}. \tag{3.2}$$

This indicates that owing to small electron effective mass ($m_e \ll m_h$) the acceptor levels turn out to be rather narrow: The width of a level is much less than its energy, $\Gamma \ll \mathcal{E}_A$. Thus the presence of acceptors in a gapless semiconductor leads to the occurrence of a narrow peak (or narrow peaks, if several species of acceptors are present) on the background of the conduction band density of states. The width of such a peak is small compared to the distance between the level and the top of the valence band.

The ground state energy of an acceptor at $m_e << m_h$, as follows from the Equation (3.1), is determined practically only by the heavy-hole mass m_h and is insensitive to the variation of electron mass m_e. But, according to the Equation (3.2), the width of the resonance level, Γ, is proportional to $(m_e/m_h)^{3/2}$. The band structure readjustment in HgCdTe when the composition x is varied or under the effect of pressure affects chiefly the gap $\mathcal{E}_g$ between the s- and p-bands (Fig.21); as a result, the electron mass m_e changes too. Therefore the position of the acceptor level $\mathcal{E}_A$ relative to the top of the valence band should remain unchanged, and the level width Γ shouls decrease when the gap $|\mathcal{E}_g|$ tends to zero. Upon inversion of the s- and p-bands, the acceptor level enters the forbidden band, and its width should tend to zero (if certainly no impurity band arises due to overlapping of the wavefunctions of adjacent acceptors).

An analysis for the case of a donor inmpurity shows that no well-defined donor levels exist in GS with $m_e << m_h$. Owing to the large density of states of the valence band, the smearing of a donor level is of the order of its energy (equal to the Bohr energy of an electron, $\mathcal{E}_{eB}$), i.e., its separation from the bottom of the conduction band. This means that the Coulomb potential well of a donor is not effective enough for an electron to be retained, and all donors are ionized even at T=0 K. It will be recalled that at electron energies $\mathcal{E} \approx \mathcal{E}_{eB}$ many-electron effects (the creation of virtual excitons) become essential. The effect of the electron-electron interaction on acceptor states is considerably weaker since the acceptor state energy $\mathcal{E}_A$ at $m_h >> m_e$ is of the order of the Bohr energy of a hole, $\mathcal{E}_{hB}$, i.e., is much larger than that of an electron.

The acceptor ground state energy $\mathcal{E}_A$ in gapless semiconductors,obtained within the framework of the one-electron approximation, varies substantially if the dispersion relation allows for the linear term in wavevector due to the exchange interaction of electrons (see Eq.(2.23)). With the exchange interaction being taken into account, the dispersion relation (see Eq.(2.27)) (and the heavy-hole mass) depends on the gap $\mathcal{E}_g$. So the energy $\mathcal{E}_A$ of the acceptor level becomes a function of $\mathcal{E}_g$ too. Calculation results for the $\mathcal{E}_A$ dependence of $\mathcal{E}_g$ (Gelmont, 1975) are presented in Figure 24 (solid line).

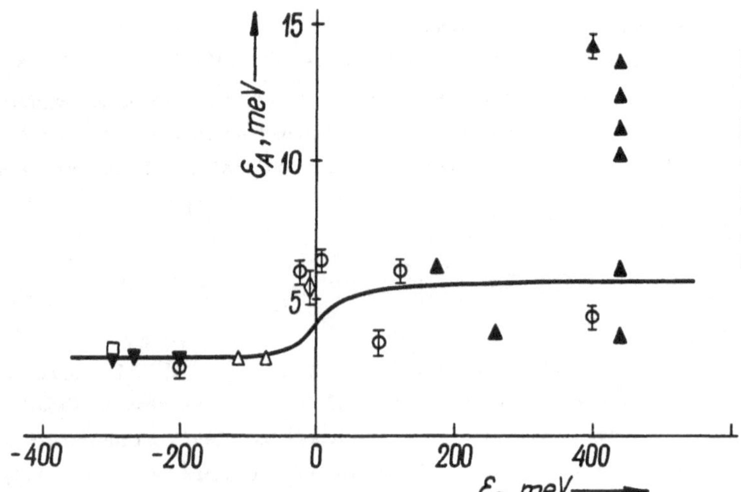

Fig.24. Dependence of the acceptor ground state energy $\mathcal{E}_A$ for HgCdTe on the gap $\mathcal{E}_g$ between the s- and p-bands. The solid line refers to a calculation, the points show experimental data by different authors .

The energy $\mathcal{E}_A$ increases monotonically by a factor of 2.5 in going from a GS with a negative gap $\mathcal{E}_g$ to a semiconductor with a positive gap $\mathcal{E}_g$. Assuming $\kappa = 20$ and $m_h = 0.4\ m_0$ for $Hg_{1-x}Cd_xTe$ crystals, the result for $\mathcal{E}_g = -300$ meV (HgTe, x=0) is $\mathcal{E}_A = 2.3$ meV, and for $\mathcal{E}_g \gtrsim 300$ meV one obtains $\mathcal{E}_A = 6$ meV.

Acceptor charge screening by free electrons at normal residual electron concentration $n \lesssim 10^{15}-10^{16}\ cm^{-3}$, when $\mathcal{E}_F \lesssim \mathcal{E}_A$, as calculation has shown, affects the quantity $\mathcal{E}_A$ little or does not at all. A somewhat stronger influence on $\mathcal{E}_A$ is exerted by screening due to peculiarities of the dielectric constant near the degeneracy point of the bands (Joos et al., 1978) (see Sect.2.3.6.). For the band parameters of HgTe, calculation gives $\mathcal{E}_A \simeq 1-2$ meV. Thus, due to peculiarities of the hole dispersion relation and the dielectric constant, the ground state energy of a hole on an acceptor in GS turns out to be appreciably less than that in a normal semiconductor with the same heavy-hole effctive mass.

We now consider some results of experimental research. For zero-gap semiconductors where the impurity levels are located on the continuous spectrum of allowed bands, the most reliable and direct method of detecting these levels is by magnetooptical investigations. According to the Fermi level position, acceptor levels may be detected as peaks in magnetoabsorption spectra either when electrons are excited from the light-hole s-band Landau levels into empty acceptor states or at transitions from the occupied acceptor states to the p conduction band Landau levels. Spectral lines due to impurities are identified from the magnetic field H dependence of the absorbed-radiation photon energy: At $H \rightarrow 0$ this dependence is extrapolated to the energy separated from the degeneracy point of the p-bands by an amount (distance) equal to the impurity activation energy in the absence of a magnetic field. Spectral lines arising from transitions to impurity levels (or from impurity levels to a band) exhibit a larger width than lines corresponding to interband transitions. This larger linewidth is due chiefly to high density of acceptor states.

Using magnetooptical methods, two acceptor activation energy values have been found for HgTe: approximately 2.2 and 0.7 meV (Bastard et al., 1974; Tuchendler et al., 1973). the first of these values is attributed to the presence of mercury vacancies that act as acceptors. This assumption is confirmed by the considerably higher impurity absorption line intensity for unannealed samples compared to that for annealed samples where the amount of mercury vacancies is not as large. The activation energy of 0.7 meV is ascribed to transitions of electrons from an occupied acceptor level to one of the empty conduction band Landau levels. Some authors attribute this acceptor level to a copper impurity.

$Hg_{1-x}Cd_xTe$ samples with a low Cd content ($x \leq 0.06$), just as HgTe, exhibit two acceptor levels whose activation energy is nearly independent of the composition of a solid solution at $x \leq 0.06$: approximately 2.5 and 0.8 meV (Guldner et al., 1977).

Thus the value of the activation energy of one of the acceptor levels in HgTe and $Hg_{1-x}Cd_xTe$ with $x \leq 0.06$ is in good agreement with theoretical predictions for $\mathcal{E}_A$ on the assumption of a Coulomb potential. We emphasize the coulombian type of potential because a number of authors have calculated the energy $\mathcal{E}_A$ for another impurity potential type, viz., a short-range potential.

The Coulomb potential is long-range; that is, at large distances r compared to the size a of a unit cell it decreases slowly with distance, as 1/r. In the immediate vicinity of a defect at $r \approx a$ the potential $V(\vec{r})$ is evidently of a complicated character. The form of the potential $V(\vec{r})$ for $r \approx a$ is determined by many factors: type of chemical bond, ionic radius, the presence of unfilled electronic shells in the case of a substitutional atom, lattice distortion (deformation) if a vacancy is present in a lattice site or if an atom is inserted in an interstice, etc.

Clearly, the potential $V(\vec{r})$ at $r \lessapprox a$ has no universal form and a calculation allowing for the aforementioned and other factors is required for each particular defect. In the general case one deals with a many-electron problem and it is not always possible even to introduce a one-electron potential $V(\vec{r})$ for the interior of a unit cell. The form of $V(\vec{r})$ is therefore modeled for $r \leq a$ in some reasonable simple way.

It is often assumed that at $r \lessapprox a$ the potential $V(\vec{r}) = V_0$, with V_0 being a constant quantity. If $V_0 < 0$ then V_0 describes a potential well for electrons; if $V_0 > 0$, then V_0 describes a potential well for holes. In many cases the correction to the bound state energy of a Coulomb well electron due to the potential V_0 is small (for p-states this correction is much less than for s-states). The bound state energy and wavefunctions are determined by the potential V_0 within the unit cell only provided the potential well is sufficiently deep, $V_0 \approx (1-10)$ meV. The behavior of $V(\vec{r})$ at large distances($r \gg a$) is immaterial.

It appears verisimilar that the potential V_0 for mercury vacancies in HgCdTe, in contrast to substitutional impurities, may turn out to be so deep that the bound state localization region will be determined by dimensions of the order of a. A number of authors have calculated the acceptor ground state energy $\mathcal{E}_A$ in HgCdTe with such a highly localized potential (Mauger et al., 1975; Bastard et al., 1977). It remains an unfortunate fact that the calculations performed prove of little utility since the formulas for $\mathcal{E}_A$ contain a great deal of unknown parameters, and the authors are therefore compelled to make rather arbitrary assumptions as to the relations between these parameters. Apart from this, in its structure, the formula for $\mathcal{E}_A$ in the localized potential model is a small difference of large quantities, a fact which renders a trustworthy quantitative treatment hopeless. At the same time a very important qualitative conclusion follows from all these calculations, which is consistent with the above calculation results for the Coulomb potential. As the gap $\mathcal{E}_g$ increases over the entire range of positive and negative values the acceptor ground state energy $\mathcal{E}_A$ increases monotonically.

We now turn to experimental results for narrow-gap HgCdTe semiconductors with $\mathcal{E}_g > 0$. When the gap between the bands is finite and the acceptor level is in the gap, the quantity $\mathcal{E}_A$ may be determined not only with the help of magnetooptical or optical investigations but also by measuring the activational temperature dependence of the Hall effect or electrical conductivity in the extrinsic conductivity region. Along with a theoretical $\mathcal{E}_A(\mathcal{E}_g)$ curve, Figure 24 presents experimental data obtained by different researchers for $Hg_{1-x}Cd_xTe$ crystals with $\mathcal{E}_g < 0$ ($x<0.16$) and $\mathcal{E}_g > 0$ ($x>0.16$). Although the experimental findings are, on the whole, in fairly good agreement with the calculation results for the Coulomb type of impurity potential, some comments must be made on the data pertaining to $Hg_{1-x}Cd_xTe$ semiconductor crystals ($\mathcal{E}_g > 0$).

The experimental points denoted by circlets have been obtained from an analysis of the temperature dependence of electrical conductivity: $\sigma \sim \exp(-\mathcal{E}_A/k_BT)$ (Arapov et al., 1979). For samples with $x \approx 0.19-0.20$, in which the gap $\mathcal{E}_g$ is about 100 meV, the electrical conductivity in the temperature interval between 10 and 40 K increases by approximately a factor of 10; it is quite probable that this may be due to excitation of holes from the acceptor level into the conduction band. However, this result should be handled with some degree of caution since the calculated $\mathcal{E}_A$ values of 3 to 6 meV only slightly exceed k_BT over the 10–40 K temperature range. In consequence, the procedure of determining $\mathcal{E}_A$ from the exponential dependence $\sigma \sim \exp(-\mathcal{E}_A/k_BT)$ cannot lay claim to being reliable enough.

For the same value $\mathcal{E}_g$ > 400 meV (corresponding to $Hg_{1-x}Cd_xTe$ crystals with $x \simeq 0.4$), a number of points are marked in Figure 24 which differ largely from each other ($3 \lessapprox \mathcal{E}_A \lessapprox 15$ meV). This data has been obtained from the exponential dependence of the Hall coefficient on temperature, $|R| \sim \exp(\mathcal{E}_A/k_BT)$, for samples with different acceptor and donor content $4 \times 10^{15} \leq N_A - N_D \leq 1 \times 10^{17}$ cm^{-3} (Scott et all., 1976). For samples with the smallest hole concentration $N_A - N_D = 4 \times 10^{15}$ cm^{-3} the calculated value $\mathcal{E}_A$ is as high as $\approx$ 15 meV, which differs markedly from the predictions of theory (solid line in Fig. 24). The point is that these samples with $x \simeq 0.4$ are highly compensated, i.e., the donor concentration is comparable with the acceptor concentration.

The difference $N_A - N_D$ may be decreased by increasing the number of donors, i.e., the degree of compensation N_D/N_A. This conclusion is borne out by an analysis of the variation of the mobility with varying $N_A - N_D$: With decrease in $N_A - N_D$ the low-temperature hole mobility, the magnitude of which is determined by the scattering of holes on charged centers, drops. An increase in mobility indicates that the total number of charged impurities, $N_A + N_D$, increases, i.e., the donors increase in number and, therefore, the compensation N_D/N_A increases too. As for highly compensated samples containing donors and acceptors in amounts that are close in magnitude ($N_A - N_D \ll N_A$), the acceptor activation energy $\mathcal{E}_A$ is appreciably larger than the discrete level activation energy calculated theoretically (Fig. 24). Due to the presence of donors, the acceptor level moves away from the valence band. This increase in activation energy $\Delta\mathcal{E}_A$ due to compensation is approximately equal to

$$\Delta \mathcal{E}_A \approx \frac{e^2 N_A^{1/3}}{\kappa} \left(\frac{N_A + N_D}{N_A - N_D}\right)^{1/3}$$

(Shklovskii and Efros, 1979).

For the samples concerned ($x \simeq 0.4$), we find (at $N_A = 5 \times 10^{16}$ cm^{-3} and $N_D/N_A = 0.9$) $\Delta \mathcal{E}_A \simeq 6$ meV. Thus the large values of activation energy $\mathcal{E}_A$ found from Hall effect measurements are not at variance (rather, if the degree of compensation of these samples is taken into account exactly, they are consistent) with the value $\mathcal{E}_A = 6$ meV which is predicted by theory for the case of a Coulomb acceptor potential.

3.4. Metal-Insulator Transitions

When discussing the energy of the bound state of electrons and holes on impurities we disregarded the effects due to the overlap of the wavefunctions of charge carriers on adjacent impurity centers, such as the broadening of and the shift of the impurity level of an isolated impurity atom and the formation of an impurity band. Indeed, these effects are immaterial when the mean distance between impurity atoms $d \approx N_i^{-1/3}$ (with N_i being the impurity concentration) considerably exceeds the dimension of localization of the wavefunction of charge carriers in the bound state - the effective Bohr radius a_B, i.e., when

$$N_i^{1/3} a_B \ll 1 \quad . \tag{3.3}$$

For a shallow impurity level in a Coulomb potential well $a_B = \kappa\hbar^2/e^2m$ (see Eq.(1.25)). In the case of an arbitrary potential at large distances from the attracting center the wavefunction of a bound state of energy $\mathcal{E}_i$ falls off exponentially ($\psi \sim \exp(-r/a_B)$) with typical scale, $a_B = \hbar/\sqrt{2m\mathcal{E}_i}$.

A semiconductor for which the condition (3.3) holds true is a weakly doped one. The density of states of an electron on an impurity has the shape of a narrow peak (this certainly holds good for holes as well), and the density of states of the conduction band is not distorted substantially by the perturbation of the impurity potential. At T = 0 K an impurity electron is localized near an impurity, and the contribution to transport phenomena is made only by conduction band electrons. At sufficiently low temperatures ($k_BT \ll \mathcal{E}_D$, for donors $\mathcal{E}_i \equiv \mathcal{E}_D$) the electron concentration increases exponentially with increase in temperature. Because of the partitioning of electronic states into localized (at $\mathcal{E} = \mathcal{E}_D$) and delocalized (at $\mathcal{E} \geqq \mathcal{E}_c$, where $\mathcal{E}_c$ is the conduction band edge) states in a slightly doped semiconductor, it is possible to determine the binding energy of an electron on a donor, $\mathcal{E}_D$, from an analysis of the temperature dependence of conductivity and Hall effect.

The case opposite to the inequality (3.3),

$$N_i^{1/3} a_B \gg 1 \quad , \tag{3.4}$$

may be realized in heavily doped semiconductors. Here the energy levels of impurity electrons are almost completely overlapped by the conduction band and no sharply defined boundary exists between the localized and delocalized states. When the inequality (3.4) is fulfilled, a volume of the order of a_B^3 contains a sufficiently large amount of impurity atoms and one has to take into account the random character of their spatial distribution. Impurity potential fluctuations generated by nonuniform impurity distribution in the crystal perturb the states of the conduction band and lead to formation of a density of states tail at energies $\mathcal{E} < \mathcal{E}_c$. Put another way, allowed states arise for electrons in the interior of the forbidden band, but the density of these states is less than the counterpart in an unperturbed conduction band.

In the vicinity of the Fermi level, which lies above the edge $\mathcal{E}_c$ of the unperturbed conduction band (since the donors are ionized in a heavily doped semiconductor even at T = 0 K), the density of states is nearly the same as that in the absence of impurities. Localized states may exist deep within the tail when $\mathcal{E} < \mathcal{E}_c$. But the electrons cannot be thermally excited from these states to the Fermi level, since the Fermi energy at large electron concentration $n = N_D$ is large (see Eq.(1.14)), and therefore do not contribute to transport phenomena. The heavy-doping condition (3.4) for finite-gap semiconductors or GS with small electron effective masses ($m \approx 10^{-2}m_o$) holds good at impurity concentrations that are not very high, at $N_D \gtrsim 10^{17}$ cm^{-3} (n-InSb, n-HgTe,

HgSe, etc.). In HgSe, as has already been mentioned, the native donor defect concentration N_D is $\gtrsim 10^{17}$ cm^{-3}. Thus, all the HgSe crystals that have been investigated so far are heavily doped in terms of the criterion (3.4).

Of topical interest for many of the HgTe and HgCdTe crystals studied is the intermediate case $N_i^{1/3}a_B \lesssim 1$. The intermediate doping case is most difficult to analyze theoretically since theory does not have the small parameter which is present both in the case of slight doping ($N_i^{1/3}a_B \ll 1$) and in the case of heavy doping ($N_i^{1/3}a_B \gg 1$). Therefore it has not been possible to ascertain the exact form of the density of states $\rho(\mathcal{E})$. Nor has it been possible to determine the energy boundary between localized and delocalized states. However, the various physical considerations based on allowance for the electron-electron interaction, Coulomb potential screening, and the degree of overlap of the wavefunctions of impurity electrons lead to the concept of the Mott transition.

This concept implies a transition from activational to metallic conductivity, a transition that is realized either by doping or under the effect of a magnetic field or pressure. The semiconductor passes from the state when the conduction occurs by electrons via impurity atoms (at T > 0 K the electrons "hop" from one impurity atom to another, which is the case of hopping conductivity) or by conduction band electrons excited from donors (in both cases the conductivity is of an activational character) to the state when the conduction occurs by conduction band electrons even at T = 0 K and is therefore of a metallic character, i.e., is independent of temperature. At some critical impurity concentration N_M, which is defined by the equation (Mott and Davis, 1971)

$$N_M^{1/3} a_B \simeq 0.25 \tag{3.5}$$

the Fermi level enters into the region of delocalized states and the conductivity becomes nonactivational, just as in metals.

The concept of the Mott transition was formulated when discussing the problem of the metal-insulator transition for solids possessing an ideal periodic crystal lattice. If one ignores the electron-electron interaction and considers an electron in the field of a periodic potential generated by all ions and the rest of the electrons, the electron wavefunction turns out to be a plane wave (just as for a free electron) having a spatially modulated amplitude with a period equal to the lattice spacing (see Eq.(1.4)). This means that in the one-electron approximation, the states of the electron in the crystal are delocalized; that is, the electron may be detected in any place of the crystal regardless of the interatomic distance, i.e., the lattice spacing.

The situation changes radically if one takes into account the electron-electron interaction. Imagine a crystal with a simple cubic lattice made up of N atoms with an electron on each outermost shell. According to band theory (see Sect. 1.2.), such a material should be a metal. In fact, two electrons with opposite spins may be present on each atomic level. In a system of N atoms that have come closer together, 2N energy levels form an allowed band which is only half filled (on each atom there is only one outer-shell electron!), the occupied and empty states adjoin each other - the material is a metal. We now take into account the electron-electron interaction. The repulsion between electrons hinders two electrons from staying near the same lattice site (let it be repeated that there is no such restriction in the one-electron approximation). For a second electron to "land" on a given site, an energy U must be spent which is needed to surmount the electron-electron repulsion. Naturally, the localization of an electron near its "host" atom is energetically more favorable.

If the atoms are separated widely enough, i.e., the lattice spacing is large, then each site has two electronic levels: $\mathcal{E}_0$ if there is one electron on the site, and $\mathcal{E}_0$ + U if there are two electrons on the site. In the state with energy $\mathcal{E}_0$, when there is one electron at each site, the crystal is an insulator. As the interatomic distance decreases the degree of overlap of the wavefunctions of adjacent electrons increases, and both levels broaden into bands, the lower band being filled, the upper band empty. As long as the bandwidth W is much less than the repulsive energy U, the electrons remain localized near their "host" atoms. At some value of lattice spacing the occupied and empty bands will touch and the electrons will lose the coupling to their "host" atoms and will become delocalized (an insulator-metal transition will occur). Thus, when the width W of the two bands under consideration is larger than the formation energy of a pair of electrons on a site, the material is a metal.

The foregoing may be put in a somewhat different way. When the concentration of atoms (and, hence, delocalized electrons) is sufficiently high the kinetic energy of delocalized electrons, which is equal to the Fermi energy $\mathcal{E}_F$ (in order of magnitude, $\mathcal{E}_F \sim W$ in the situation considered), is larger than the mean electron interaction energy U, and the one-electron approximation proves to be well-founded.

Thus the electron-electron interaction is a major factor that precludes the conservation of delocalized states of an electron in a periodic potential field when the interatomic distances are large. The physical causes leading to the Mott transition hold good also in the more general case of disordered condensed systems, viz., amorphous materials, glasses, etc. We are interested in the disordered system of impurity atoms in an ideal crystal which perturb the lattice periodicity and give rise to a random impurity potential.

Two situations must be distinguished. Let a random potential be generated by like impurity atoms or ions that are randomly arranged in the crystal. Such a potential may be represented as a set of potential wells of equal depth V but with arbitrary spacing. Apparently this model is fairly good for describing a semiconductor with one species of impurities, either donors or acceptors. The width W of the impurity band arising from the overlap of the wavefunctions of adjacent impurity atoms is a random quantity, but its characteristic value is determined by the mean distance between the impurity atoms, $\bar{d} \approx N_i^{-1/3}$. Therefore the condition for the Mott transition (3.5) remains valid for such a potential.

Assume now that the impurity atoms are spaced equal distance apart, i. e., form a regular lattice, but the electronic levels of the atoms of this lattice differ in energy (Anderson, 1958). The random potential is an assembly of periodically arranged potential wells of different depth. The Anderson model is applicable to a compensated semiconductor containing both neutral impurity atoms and ionized donors and acceptors.

When the impurity distances are large the wavefunction of an electron on an impurity atom is localized even with neglect of the electron-electron interaction. What hinders delocalization in this model is not the electron-electron repulsion but a spread ΔV in values of the random potential V. It is only when the distance between the impurity centers decreases (as a result of an increase in impurity concentration) to the extent that the width W of the impurity band produced becomes comparable with or exceeds ΔV that an energy band corresponding to delocalized states appears in the electron spectrum. The electrons whose energy is outside this interval remain localized, although the share of localized states diminishes with increasing impurity concentration. The transition to metallic conductivity in the model described - the Anderson transition - occurs evidently when the Fermi level enters into the energy band of delocalized states.

We have considered two metal-insulator transition types- the Mott transition and the Anderson transition. The fundamental difference between them is that the Anderson transition arises from the randomness of the system, whereas the Mott transition occurs in an ideal periodic structure. In addition, the Mott transition arises as a result of the electron-electron interaction, whereas the Anderson transition relates to the case of one-electron motion. In the case of noncompensated semiconductors (one species of impurities) a criterion for the transition of a disordered system of impurities to metallic conductivity, which should be called the Mott-Anderson transition, may be the Equation (3.5). For highly compensated semiconductors with $N_D \simeq N_A$, theory provides no sufficiently convincing and experimentally verified quantitative criteria, and only qualitative estimates based on reasonable physical considerations can be used.

It has not been possible to make reliable quantitative estimates of the density of states at impurity concentrations N_i near the critical concentration N_M (3.5) corresponding to the Mott transition. However, it seems to be clear qualitatively that the energy boundary separating localized and delocalized states (the mobility edge, according to Anderson's terminology) at impurity concentrations $N_i \simeq N_M$ should pass near the Fermi level at T = 0 K. Therefore electrons(holes) can readily be excited thermally from localized states into delocalized ones and contribute to charge transport phenomena.

3.5. Mott Transition in n-Type Crystals

In a normal semiconductor, when the temperature is lowered and the gap $\mathcal{E}_g$ between the bands or the gap $\mathcal{E}_D(\mathcal{E}_A)$ between the donor (acceptor) levels and the conduction (valence) band increases, the electron (hole) concentration determined by the Boltzmann statistics should decrease, i.e., the electrons (holes) should be frozen out from the delocalized states where they participate in charge transport. We consider the problem of electron freeze-out in narrow-gap and zero-gap n-type HgCdTe semiconductors ($N_D \gg N_A$).

The critical concentration N_M (3.5) at which the transition from activational to metallic conductivity occurs may be represented as

$$N_M = 10^{23} \left(\frac{1}{\kappa} \frac{m}{m_0} \right)^3 \text{cm}^{-3} . \qquad (3.6)$$

Here we have used the expression for the Bohr radius a_B in terms of dielectric constant κ and effective mass m. In the two-band approximation, the electron effective mass m_e^* at the bottom of the conduction band is expressed (see Eq. (2.20)) in terms of the quantities P and $\mathcal{E}_g$. Allowing for this relation yields for the donor concentration at which the Mott transition sets in

$$N_M = (\frac{3e^2}{16\kappa P^2} \quad |\mathcal{E}_g|) \ . \tag{3.7}$$

This equation tells us that the concentration N_M diminishes very rapidly with decreasing gap $|\mathcal{E}_g|$. Thus for the gap $|\mathcal{E}_g|$ = 300 meV, which corresponds to the solid solution $Hg_{0.7}Cd_{0.3}Te$ and GS HgTe, at $\kappa = 20$ and $P = 8 \times 10^{-8}$ eV cm, $N_M = 2 \times 10^{14}$ cm^{-3}. For a $Hg_{0.8}Cd_{0.2}Te$ semiconductor with $\mathcal{E}_g \simeq 0.08$ eV at T = 4.2 K, $N_M \simeq 2 \times 10^{13}$ cm^{-3}. In all of the HgTe and HgCdTe crystals obtained to date the donor concentration is larger than 10^{14} cm^{-3}, i.e., $N_D \gtrapprox N_M$. This means that the localized and delocalized states are not separated, and no freeze-out of conduction band electrons into donors is possible. The well established fact that no electrons are frozen out into donors with decrease in temperature in either narrow-gap crystals n-$Hg_{0.8}Cd_{0.2}Te$ or GS HgTe thus becomes understandable. Incidentally, no freeze-out into donors has been detected in α-Sn, either.

Let it be recalled once again that in GS an electron cannot be retained by a donor having a Coulomb-type potential (there are no bound states of an electron on a donor). In other words, the wavefunction localization radius of an electron on a donor is infinite and all donors are ionized however small the donor concentration may be. If the donor potential is assumed to be localized within a unit cell, theory admits the possibility of there existing localized electrons and sharp donor levels if only the probability of an electron passing from a donor level to valence band continuous-spectrum states of the same energy is small. Unfortunately, for all GS $Hg_{1-x}Cd_xTe$ ($x <$ 0.16) and for α-Sn the minimum donor concentration N_D is larger than the quantity N_M which is given by the formula (3.5) for shallow donor levels with $\mathcal{E}_D \simeq 1$ meV (to perform estimates one has to use the formula for the Bohr radius $a_{eB} = \hbar/\sqrt{2m_e \mathcal{E}_D}$). Therefore the cause for the smearing of donor levels and, consequently, the absence of the electron freeze-out in GS may currently not be stated unambiguously. This smearing may result from both the overlap of the electron wavefunctions of neighbouring donors (concentration-induced broadening of donor levels) and the finite probability of a transition of an electron from a donor level to delocalized valence band states of the same energy (natural broadening of levels).

Thus up till now it has not been possible to detect localized states of electrons on donors in n-type gapless and narrow-gap semiconductors because of a large donor impurity content $N_D > N_M$. But cannot one make an attempt to reveal localized states of electrons in a high magnetic field?

Indeed, in a strong magnetic field the electron spectrum is quantized (Sect. 2.3.7.), the parabolic-band edge being displaced by $\hbar\omega_{ce}/2$ (Fig. 18a)). If the intensity of the magnetic field H reaches such a magnitude that the displacement of the conduction band edge is larger than the ground bound state energy of an electron in a Coulomb donor well, i.e.,

$$\frac{\hbar\omega_{ce}}{2} > \mathcal{E}_D , \tag{3.8}$$

a gap arises between the lower Landau level of the conduction band and the donor level and at temperatures $k_B T < \mathcal{E}_D$ the electrons are localized on donors. Allowing for $\mathcal{E}_D$ =

$m_e e^4/(2\hbar^2\kappa^2) = e^2/(2\kappa a_{eB})$, the inequality (3.8) may be written in the form

$$L < a_{eB} \qquad (3.9)$$

with $L = (\hbar/eH)^{1/2}$ being the magnetic length (2.46), which is the typical dimension of the region of localization of the electron wavefunction in the direction perpendicular to $\vec{H}$, i.e., $L \approx a_{eB}^{\perp}$. The occurrence of a gap between the donor states and the conduction band in a magnetic field is due to the magnetic field leading to compression of the electron wavefunction; this compression is particularly substantial in the direction perpendicular to $\vec{H}$ (Yafet et al., 1956). As a result of a decrease in Bohr radius $a_{eB}^{\perp}$, the overlap of the electron wavefunctions of adjacent donors decreases and the impurity band becomes narrower, which promotes the occurrence of a gap between the donor states and the conduction band.

The condition for the Mott transition may be written similar to Equation (3.5) as

$$\left[N_D (a_{eB}^{\perp})^2 \, a_{eB}^{\parallel} \right]^{1/3} \approx 0.25 \;, \qquad (3.10)$$

where $a_{eB}^{\parallel}$ is the dimension of the region of localization of the electron wavefunction in the direction parallel to $\vec{H}$. The minimum magnetic field H_M in which localized states of electrons appear on donors is, according to the Equation (3.5), determined by

$$H_M = \frac{\kappa}{m_e} \left(\frac{4\hbar}{e} \right) N_D \qquad (3.11)$$

(we set $a_{eB}^{\perp} = L$, $a_{eB}^{\parallel} = a_{eB}$).

The expression (3.10) has not been obtained rigorously and is a semiempirical generalization of the formula (3.5). At the same time it has turned out to be in good agreement with experimental data. Formation of localized states under the effect of a magnetic field leads to an increase in the Hall coefficient $|R| = 1/en$ as well as the transverse ρ_{xx} and longitudinal ρ_{zz} magnetoresistance owing to magnetic freeze-out of originally delocalized electrons. This effect was repeatedly observed on n-InSb samples with different impurity content $N_D > N_M$. The value of critical field H_M above which the Hall coefficient $|R(H)|$ displays a sharp increase that reflects the decrease in the concentration of delocalized conduction band electrons agrees well with the magnitude given by the expression (3.11) (Neuringer, 1968; Robert et al., 1980).

Studies of this effect on n-type $Hg_{1-x}Cd_xTe$ ($0.18 \le x \le 0.35$) with $n = 10^{14}$ to 10^{16} cm^{-3} have shown the behavior of galvanomagnetic coefficients to differ from that observed on n-InSb samples. In quantizing magnetic fields ($\hbar\omega_{ce} \gg k_BT$) the longitudinal and the transverse magnetoresistance increase, just as in InSb, severely with increasing field H but the Hall coefficient depends weakly and nonmonotonically on H. This behavior of R(H) can be explained if one takes into account the contribution to conductivity by incompletely localized impurity state electrons which are able to move over the impurity donor band under the effect of an electric field. Although the mobility of these electrons is considerably less than that of band electrons, the concentration of

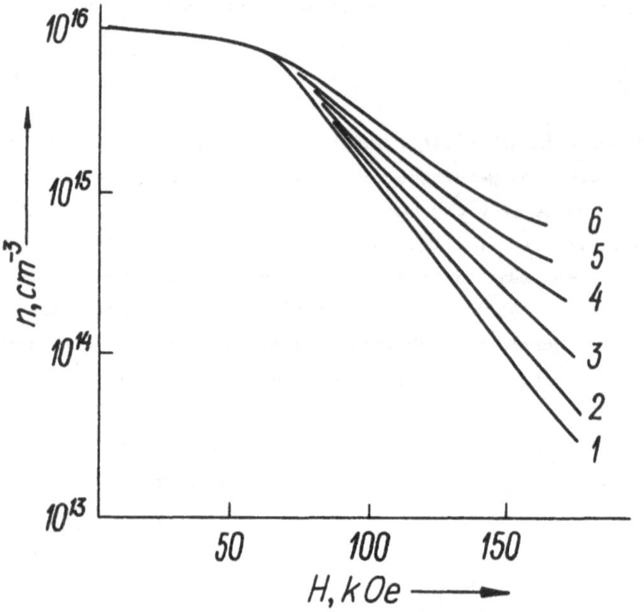

Fig.25. Magnetic field dependence of the conduction band electron concentration n calculated from experimental data using Eq.(3.12) for a $Hg_{0.75}Cd_{0.25}Te$ sample at different temperatures, K: (1) 2, (2) 3.4, (3) 4.2, (4) 4.9, (5) 5.6, (6) 6.5

the latter is related to the Hall coefficient by an equation that is different from the simple n = 1/eR (Robert et al., 1980):

$$n = \frac{RH^2}{e(R^2H^2 + \rho_{xx}^2)} \quad . \tag{3.12}$$

Figure 25 illustrates the effect of electron-freeze-out in a $Hg_{1-x}Cd_xTe$ (x = 0.25) sample with $n = 1 \times 10^{16} cm^{-3}$ at 4.2 K (Raymond et al., 1981). According to the Equation (3.5), the critical concentration of the Mott transition, N_M, is $\approx 1 \times 10^{14} cm^{-3}$. Therefore, in the absence of a magnetic field all the electrons are in the conduction band. In fields H ≈70 kOe a gap forms between the conduction and valence bands which increases with field, and the electron concentration n in the conduction band starts decreasing.

Figure 26 presents the magnetic field dependence of the activation energy of electrons from donors, $\varepsilon_D(H)$, for a $Hg_{0.66}Cd_{0.34}Te$ sample with $N_D = 4.5 \times 10^{14} cm^{-3}$ (Raymond et al., 1980), which is somewhat lower than the value $N_M \approx 10^{15} cm^{-3}$ for a crystal with such a cadmium content. As is seen from the figure, the electron activation energy ε_D increases with magnetic field in complete accord with theory's predictions.

3.6. The Effect of Compensation on the Mott Transition

Of interest are the features peculiar to the effect of the magnetic freeze-out of electrons into donors and acceptors in compensated HgCdTe crystals ($N_D \approx N_A$). We start by considering n-type semiconductors. Hall effect and magnetoresistance investigations on n-$Hg_{0.8}Cd_{0.2}Te$ crystals with electron concentration $n \le 1 \times 10^{15} cm^{-3}$ and with the degree of compensation $(N_A/N_D) \approx 0.9$ have shown that the freeze-out of electrons from the conduction band (a sharp decrease in n with increasing H) begins in magnetic fields

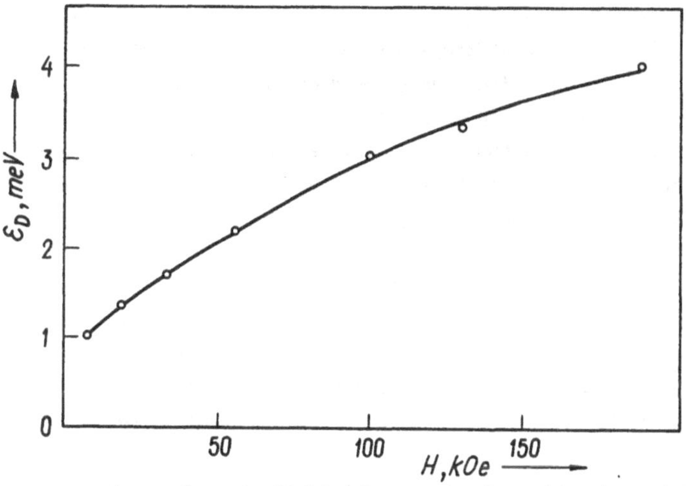

Fig.26 Magnetic field dependence of electron activation energy ε_D for a $Hg_{0.66}Cd_{0.34}Te$ crystal

H that are an order of magnitude lower than those which follow from the Mott transition criterion (3.10). It turns out to be on account of the noticeable compensation of samples that the magnetic freeze-out of electrons takes place in fields that are weaker than the critical field H_M (3.11) (Arapov et al., 1983)

In fact, the condition for the Mott transition (3.5), as well as its modification (3.10) holds only when the sample contains impurities of the same species (say, donors) or at least the degree of compensation is weak enough for the electron concentration n to be slightly short of the ionized donor concentration.

Let the donor concentration N_D of the sample be larger than the critical concentration of the Mott transition, N_M, i.e., all donors be ionized and all electrons be delocalized, $n = N_D$. When acceptors are added, i.e., with increase in the degree of compensation N_A/N_D, there is a decrease in the concentration of delocalized electrons $n = N_D - N_A$ (N_A electrons are captured by acceptors) and, hence, in their Fermi energy. In addition, as the concentration n is decreased the screening of charged impurities by electrons is weakened (the screening radius r_0 of the impurity potential increases)[19]. It is shown in the theory of heavily doped semiconductors that the larger the total ionized impurity atom concentration and the screening radius, the deeper the spatial impurity potential fluctuations that distort the conduction band edge and produce a potential relief[20].

Thus with increasing N_A/N_D at fixed concentration N_D the Fermi energy decreases and the depth of the potential wells, i.e., the potential energy of electrons, increases. This means that as the compensation is increased the tendency of electrons for passing from delocalized to localized states enhances. Clearly, the critical donor concentration corresponding to the transition to delocalized states should in this case be higher

19) The screening radius for a degenerate electron gas is

$$r_o = \frac{\hbar}{2e}\left(\frac{\kappa}{m}\right)^{1/2}\left(\frac{\pi}{3n}\right)^{1/6}.$$

20) Potential fluctuations of a scale larger than r_0 are completely screened by electrons and a major contribution to the potential is made by fluctuations with a scale of the order of r_o. Since a fluctuation in a volume of linear dimensions R decreases or increases the energy of electron by the amount $\gamma(R)=(e^2/\kappa)(N_a+N_D)^{1/2}R^{1/2}$, fluctuations of dimensions $R<r_o$ evidently make a smaller contribution than those of dimensions of the order of r_o.

than that following from the criterion (3.5).

The magnetic field leads to localization of the electron wavefunction in the region of typical dimension $L = (\hbar/eH)^{1/2}$ in the direction perpendicular to field. In turn, this brings about an increase in the electron density of states near the bottom of each Landau subband (Fig.18b)). In the ultraquantum limit, i.e., at $\hbar\omega_{ce} \gg \mathcal{E}_F$, when only the lowest-lying Landau subband is filled, the Fermi level shifts down on the energy scale toward the bottom of the lower Landau subband, i.e., toward the bottom of the conduction band.

As has been already stated previously, the Fermi level in compensated semiconductors lies lower than in noncompensated ones and the fluctuations of the impurity potential are larger. Therefore the threshold magnetic field needed to freeze out the conduction band electrons into the localized states located in the potential relief wells should be lower than that following from the condition (3.11) for a noncompensated semiconductor ($N_A = 0$).

In the case of highly compensated n-$Hg_{0.8}Cd_{0.2}Te$ crystals with deep potential wells, the effect of the magnetic field reduces mainly to the displacement (lowering) of the Fermi level into the region of a larger density of electron states. The effect of the compression of the wavefunctions of donor electrons in samples with such deep potential wells does not suffice to separate impurity states from the conduction band; to this end magnetic fields are required that are stronger than those used in experiments. The occurrence of an activation energy in a magnetic field as revealed by Hall effect and magnetoresistance investigations corresponds to the displacement of the Fermi level from the region of delocalized states into localized states of electrons in potential wells.

We now turn to p-type zero-gap semiconductors. The acceptor levels in such semiconductors are not sharply defined since they lie in the region of the continuous conduction band spectrum, and therefore no strictly localized holes exist on acceptors. However, the natural level Γ broadening due to the finite probability of transition to conduction band states of the same energy is much less than the ground state energy of a hole on an acceptor, $\Gamma \ll \mathcal{E}_A$. Therefore the delocalized states of valence band holes and the quasilocalized[21)] states on an impurity center are separated by an energy gap. With increase in acceptor concentration there is an increase in the impurity band width (concentration-induced broadening) which depends on the degree of overlap of acceptor wavefunctions. So the critical concentration N_M corresponding to the Mott transition (the concentration at which the energy gap between the valence band and the acceptor states disappears) retains its sense for both zero- and finite-gap semiconductors. In this case, too, the value of N_M is determined by the Equation (3.5), in which a_B implies the Bohr radius a_{hB} of a heavy hole.

In order of magnitude, the value of the critical concentration N_M for GS $Hg_{1-x}Cd_xTe$ with $x < 0.16$ should be the same as that for narrow-gap semiconductors with $x > 0.16$, for the heavy-hole effective masses $m_h \approx 0.4\ m_o$ are about the same for solid solutions with different cadmium content. Since the energy of the level of an isolated acceptor,

21) The term "quasilocalized" implies here that it is necessary to take into account the finite contribution to the conductivity by the charge carriers that are in these states.

$\mathcal{E}_A$, in gapless semiconductors is somewhat less than that in narrow-gap semiconductors (for HgTe with $x = 0$ $\mathcal{E}_A \simeq 2$ meV, for crystals with $x > 0.16$ $\mathcal{E}_A \simeq (5$ to $6)$ meV), the critical concentration in the former should be less (recall that $a_{hB} = \hbar/\sqrt{2m_h\mathcal{E}_A}$): For zero-gap HgTe crystals $N_M \approx 5 \times 10^{16}$ cm^{-3}, for a narrow-gap semiconductor $N_M \approx 2 \times 10^{17}$ cm^{-3}.

As already noted at the end of Section 3.4., the energy boundary between localized and delocalized states at impurity concentrations $N_i \approx N_M$ should pass near the Fermi level at $T = 0$ K. It is therefore to be expected that for $N_A > N_M$, say $N_A \gtrapprox 5N_M$, the Fermi level at $T = 0$ K is located in the valence band ($\mathcal{E}_F < 0$) and, consequently, there are no electrons in the conduction band: $n = 0$. When $N_A < N_M$, say $N_A \lessapprox 0.2$ N_M, the Fermi level lies in the conduction band and the electron concentration n is determined by the concentration of completely ionized donors: $n = N_D$.

An analysis of the measured temperature dependence of conductivity and Hall effect for p-HgTe crystals with hole concentrations $10^{15} < (N_A - N_D) \leq 10^{19}$cm^{-3} shows that the Mott transition occurs at $(N_A - N_D) \approx 1 \times 10^{17}$cm^{-3} (Tsidilkovski et al., 1985). This agrees with the foregoing theoretical estimate of the critical concentration N_M. The theoretical evaluation of the quantity N_M according to the formula (3.5) may therefore be regarded as quite reliable for GS as well. For $Hg_{1-x}Cd_xTe$ zero-gap semiconductors with $x \leq 0.16$ the acceptor state energy $\mathcal{E}_A$ varies from 2 to $\approx$ (5-10) meV and, correspondingly, N_M assumes values between 5×10^{16} and 2×10^{17}cm^{-3}.

We have states in the foregoing that an increase in compensation shifts the Mott transition boundary toward larger values of N_M compared to those given by the condition (3.5). The effect of compensation is particularly substantial for heavily doped p-type gapless semiconductors. Under heavy doping conditions all impurities are ionized. Spatial impurity distribution nonuniformities lead to formation of a valence band density of states tail which penetrates into the conduction band; that is, an effective overlap of the conduction and valence band arises. In the conduction band, too, a density of states tail forms which penetrates into the interior of the valence band. However, because of the small electron effective mass $m_e << m_h$, the number of states in the conduction band tail is insignificant.

The Fermi level position is determined by the hole concentration $p = N_A - N_D + n$ (see Eq.(1.28)) and the number of states $\tilde{p}$ in the valence band tail. The quantity $\tilde{p}$ depends on the amplitude of the random potential γ, whose mean value for a heavily doped and highly compensated semiconductor is

$$\bar{\gamma} \simeq \frac{e^2}{\kappa} \frac{(N_A + N_D)^{2/3}}{p^{1/3}} . \tag{3.13}$$

When the degree of compensation is low ($N_D/N_A < 0.5$),

$$\bar{\gamma} = \frac{e^2}{\kappa r_o} (4\pi N_A r_o^3)^{1/2} , \tag{3.14}$$

with r_o being the screening radius. As the degree of compensation N_D/N_A is decreased, the hole concentration p decreases and the depth of the potential wells, $\bar{\gamma}$, increases and, consequently, the number of states in the valence band tail, $\tilde{p}$, increases, too. It is obvious that at $\tilde{p} > p$ the Fermi level enters into the conduction band and, in virtue of the neutrality Equation (1.28) $p = N_A - N_D + n$, electrons inevitably appear in p-type zero-gap semiconductors. At a sufficiently high degree of compensation ($p \ll N_A + N_D$) the condition $p \gg \tilde{p}$ may be attained. In that case the Fermi level penetrates into the conduction band to a depth of the order of $\bar{\gamma}: \mathcal{E}_F \approx \bar{\gamma}$, the electron concentration becoming equal to $n = (2^{3/2}/3\pi^2)(N_A+N_D)/(pa_{eB})^{1/2}$.

Figure 27 graphically demonstrates the shape of the density of states and the Fermi level position in p-type GS as a function of the level of doping and the degree of compensation N_D/N_A. Figure 27a) refers to the case of weak doping ($N_A \ll N_M$) and weak compensation ($N_D \ll N_A$). The Fermi level position is determined by the electron concentration n equal to the concentration N_D of donors which are completely ionized. Figure 27b) presents the case of weak doping ($N_A \ll N_M$) and moderate compensation ($N_D \lesssim N_A$). This situation is important in practice and requires particular attention.

In the absence of acceptors, i.e., when $N_A = 0$, the Fermi level lies in the conduction band and its position at T = 0 K is determined only by the donor concentration according to the Equation (1.14): $\mathcal{E}_F = \hbar^2/2m_e(3\pi^2 n)^{2/3}$.

If $N_A > N_D$, the number of cunduction band states, n_0, over the energy range $0 \le \mathcal{E} \le \mathcal{E}_A$ is equal to $n_o = (1/3\pi^2)(2m_o\mathcal{E}_A/\hbar^2)^{3/2}$, whence it follows that $\mathcal{E}_A = (\hbar^2/2m_o)(3\pi^2 n_o)^{2/3}$. When the donor content N_D is less than the number of states, n_o, in the conduction band below the acceptor level, the Fermi energy at T = 0 K is determined by the formula (1.14). Since $\mathcal{E}_F < \mathcal{E}_A$, all acceptors in this situation remain neutral: The number of ionized acceptors, N_A^-, is equal to zero. This is what redically distinguishes a p-type GS from a normal compensated semiconductor in which at T = 0 K the concentration N_A^- is equal to N_D and the Fermi level coincides with the acceptor level at an arbitrary small donor concentration. In GS the compensation sets in only at $N_D \ge n_o$, when electrons can be captured from donors by acceptors. The electroneutrality Equation (1.28) in this case has the form

$$N_A^- = N_D - n_o \quad (n = n_o). \tag{3.15}$$

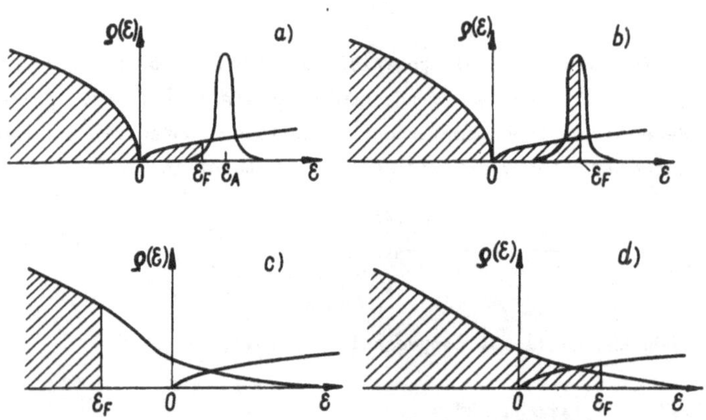

Fig.27. Density of states and Fermi level position for a p-type GS

A comparison of the above expressions for $\mathcal{E}_F$ and $\mathcal{E}_A$ shows that

$$\mathcal{E}_F = \mathcal{E}_A . \tag{3.16}$$

For HgTe ($m_e = 0.03\ m_o$, $\mathcal{E}_A \simeq 2$ meV) we find $n_o = 2 \times 10^{15} cm^{-3}$. For GS HgCdTe n_o diminishes with decreasing gap $|\mathcal{E}_g|$: $n_o \sim |\mathcal{E}_g|^{3/2}$. Since the minimum donor concentration $N_D \gtrsim 10^{15} cm^{-3}$ in the GS investigated to date, N_D is practically always larger than or equal to n_o, and the Fermi level at low temperatures is stabilized according to the Equation (3.16) or, as one sometimes says, "frozen" into the acceptor states. In a quantizing magnetic field or under hydrostatic pressure the electron spectrum and the density of states change. Apart from this, the quantity n_o may decrease appreciably, but the Fermi level remains fixed as there are a great deal of neutral acceptors capable of accepting all n_o electrons.

Rigid fixation of the Fermi level is possible only on an acceptor level of vanishingly small width $\Gamma \to 0$. Under real conditions, when the quantity Γ is finite or even comparable with $\mathcal{E}_A$, the Fermi level is not fixed strictly at a certain energy $\bar{\mathcal{E}}$. However, its position is very weakly sensitive to the variation of the band parameters of electrons, say the effective mass, if only the density of states $\rho_c(\bar{\mathcal{E}})$ at the conduction band energy $\bar{\mathcal{E}}$ is much less than the density of acceptor states $\rho_A(\bar{\mathcal{E}})$. This property of the GS is a general one and does not depend on the nature of those states which overlap with the conduction band and are a reservoir for electrons. For example, the Fermi energy $\mathcal{E}_F$ depends very weakly on electron effective mass m_e both for the fluctuational band overlap model of an intrinsic GS and for a semimetal with $m_h \gg m_e$.

Figure 27c) refers to a heavily doped ($N_A \gg N_M$) and weakly compensated ($N_D \ll N_A$) gapless semiconductor. The Fermi level is determined by the hole concentration $p \simeq N_A$ and is located in the valence band. The states in the band tail are unoccupied. Finally, Figure 27d) depicts schematically the density of states of a heavily doped ($N_A \gg N_M$) and highly compensated ($N_A - N_D \ll N_A$) GS. The Fermi level lies deep within the density of states tail of the valence band: $\mathcal{E}_F \simeq \bar{\gamma}$.

3.7. An "Anomaly" in the Temperature Dependence of Electron Concentration

Gapless semiconductors display one important feature: In the presence of acceptors the conduction band electron concentration and the Fermi energy vary "anomalously" with temperature if one compares these changes with the corresponding dependence for a normal semiconductor (Gelmont and Dyakonov, 1972). In a usual semiconductor the band electron concentration increases exponentially with temperature as a result of the excitation of the band electrons from their valence band or from the donor levels (curve 3 in Fig. 28 refers to intrinsic conductivity). In an intrinsic GS the electron concentration exhibits a power-type temperature dependence: $n \sim T^{3/2}$ (curve 2 in Fig. 28). In GS all donors are ionized at T = 0 K and there are $n = N_D$ electrons in the conduction band[22].

22) In the absence of donors in a GS at T=0 K the concentration n is equal to zero, but electrons and holes emerge at temperatures as low as desired, the intrinsic conductivity arising at lower temperatures than the extrinsic conductivity due to the capture of electrons by acceptors.

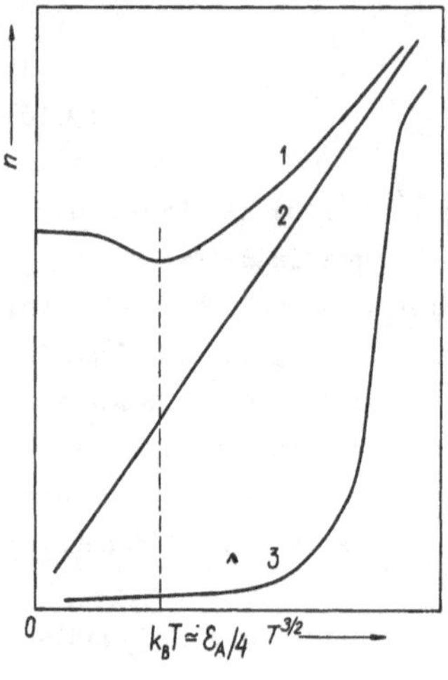

Fig.28. Temperature dependence of band electron concentration: (1) GS with acceptors, (2) intrinsic GS, (3) intrinsic semiconductor

For the acceptors to be noticeably ionized, i.e., to capture conduction band electrons of energies less than the acceptor ground state energy $\mathcal{E}_A$, it is necessary to elevate the temperature by such an amount that the energy k_BT is comparable in order of magnitude with $\mathcal{E}_A$. If the acceptors are few in number and the acceptor level is smeared weakly, the electrons captured by the acceptors become practically bound. With increase in temperature the band electron concentration should therefore decrease in some temperature interval. As the temperature is raised further, when the acceptors become depleted of vacant sites (all acceptors are ionized), the electrons begin to be excited from the valence band into conduction band states, the band electron concentration increases, i.e., intrinsic conduction sets in.

Thus the band electron concentration n in a GS containing acceptors should vary with increasing temperature in an unusual manner: It first decreases and then starts increasing (curve 1 in Fig. 28).

The dependence n(T) may be calculated using the electroneutrality equation

$$n + N_A^- = p + N_D \quad ,$$

where the ionized acceptor concentration (see Eq.(1.30)) is

$$N_A^- = N_A \left[1 + 4\exp\left(\frac{\mathcal{E}_A - \mathcal{E}_F}{k_BT} \right) \right]^{-1} . \tag{3.17}$$

The factor 4 appears as a consequence of the four-fold spin degeneracy of the acceptor level split off from the degenerate valence band. For the low-temperature region, $k_BT \ll \mathcal{E}_A$, the contribution of holes p in the Equation (1.28) may be neglected, and the Equations (3.15) and (3.16) yield

$$\mathcal{E}_F = \mathcal{E}_A - k_BT \ln \left(\frac{N_A}{N_D - n_o} - 1 \right). \tag{3.18}$$

This equation says that the Fermi energy $\mathcal{E}_F$ decreases linearly with temperature if

$$N_A > 2(N_D - n_o) \ . \tag{3.19}$$

In GS with $|\mathcal{E}_g| < 100$ meV N_D is normally much larger than n_o, and the condition (3.19) means that the acceptor level is less than half filled with electrons. In the low-temperature region considered the electron gas is degenerate, $n \sim \mathcal{E}_F^{3/2}$, and when the condition (3.19) is fulfilled the electron concentration should decrease with temperature. The Fermi level in this case lowers too, thereby promoting the thermal excitation of electrons from the valence band and the appearance of holes.

It is not possible to obtain an analytic expression for n(T) and $\mathcal{E}_F(T)$ over the entire temperature interval, but the physical cause for the nonmonotonic n(T) dependence is evident. This results from the competition between the process of electron capture by neutral acceptors and the process of thermal excitation of electrons from the valence band. Since N_D = const, it follows from the Equation (1.28) that

$$\frac{dn}{dT} = \frac{dp}{dT} - \frac{dN_A^-}{dT} \,. \tag{3.20}$$

At low temperatures $dp/dT \simeq 0$ and the first process dominates. At high temperatures, $k_BT \simeq \mathcal{E}_A$, all acceptors are ionized (depletion region), $dN_A^-/dT \simeq 0$, and according to the Equation (3.20), $dn/dT > 0$. If the condition (3.19) is not fulfilled, say $N_D \approx N_A$ (strong compensation), and the acceptor level at T = 0 K is practically filled with electrons, the concentration n and the Fermi energy $\mathcal{E}_F$ should increase monotonically with temperature.

In the case when $N_D < n_o$ and $\mathcal{E}_F < \mathcal{E}_A$, the character of the n(T) dependence is determined by the position of the Fermi level $\mathcal{E}_F$ relative to the acceptor level $\mathcal{E}_A$. An analysis of the Equation (3.20) permits establishing the lower bound of $\mathcal{E}_F$ values above which the n(T) dependence is nonmonotonic. At low temperatures $dp/dT \sim \exp(-\mathcal{E}_F/k_BT)$ (the hole gas is nondegenerate because of the large effective mass) and

$$\frac{dN_A^-}{dT} \sim \exp\left(-\frac{\mathcal{E}_A - \mathcal{E}_F}{k_BT}\right) ,$$

i.e., both terms on the right-hand side of the Equation (3.20) are exponentially small. It is clear, however, that $dn/dT < 0$ at $T \to 0$ if

$$\mathcal{E}_F > \frac{\mathcal{E}_A}{2} \quad \left(\text{ or } N_D > \frac{n_o}{2^{2/3}}\right) , \tag{3.21}$$

i.e., if the Fermi level is closer to the acceptor level than to the band edge. When the number of donors is comparatively small, so that $\mathcal{E}_F < \mathcal{E}_A/2$, the n(T) relation is monotonic.

Numerical computations have shown that both when $\mathcal{E}_F \simeq \mathcal{E}_A$ and when $\mathcal{E}_F > \mathcal{E}_A/2$ the position of the minimum n(T) is practically independent of the acceptor concentration N_A: $k_BT_{min} \simeq (0.2 \text{ to } 0.25)\mathcal{E}_A$ (Fig. 28). However, the depth of the minimum increases appreciably with N_A. Thus at $T = T_{min}$ the electron concentration n decreases by not more than 20 to 30% for $N_A = 10n_o$ and by nearly a factor of 3 for $N_A = 100n_o$.

The predicted effect of the nonmonotonic temperature dependence of electron concentration was rather difficult to detect experimentally. It would seem that the larger the concentration of acceptors ($N_A \gg N_D$) serving as a reservoir for electrons, the more favorable the conditions for observing the effect should be. However, an impurity band forms at large N_A, whereas the effect of the nonmonotonicity of n(T) has been predicted for a sharp acceptor level. It is clear that due to the broadening of the level the nonmonotonicity of n(T) may weaken appreciably. Calculations have shown that the nonmonotonicity vanishes altogether when the impurity band width becomes comparable with the energy of the level $\mathcal{E}_A$.

Experimental difficulties in observing the effect come also from the fact that, as a rule, several charge carrier types contribute to transport phenomena in GS at low temperatures: conduction band electrons, acceptor band holes and sometimes also valence band holes, charge carriers of layers adjacent to the surface. And yet it has been possible, by separately determining the concentration and mobilities of band electrons and low-mobility charge carriers, to reveal the nonmonotonic n(T) dependence from conductivity and Hall effect measuremtens on $Hg_{0.9}Cd_{0.1}Te$ samples with $N_A \gtrsim 10^{17}$ cm^{-3} (Giriat et al., 1976). The weakly marked minimum has been found to lie at $T \simeq 5$ K. In spite of the fact that at $N \approx 10^{17} cm^{-3}$ the acceptor level may certainly not be regarded as sharp, the experimental value of T_{min} is in reasonable agreement with theoretical predictions for a well-defined level; according to theory, $k_B T_{min} \simeq (0.2$ to $0.25)\mathcal{E}_A$ with $\mathcal{E}_A = (2.4 \pm 0.3)$ meV.

3.8. Freeze-Out of Electrons into Acceptors under the Effect of a Magnetic Field and Pressure

Section 2.3.7. treated the problem concerning the electron spectrum reconstruction of a pure gapless semiconductor in a magnetic field. A sufficiently strong magnetic field is capable of opening up a gap between the conduction and valence bands. But how does the magnetic field affect the position of impurity levels?

In sufficiently strong fields for which the condition (3.8) (or 3.9) holds, the impurity ground state energy increases with field as a result of the compression of the electron wavefunctions, i.e., the gap between the impurity level and the band from which it is split off increases. For $\mathcal{E}_A = 2$ meV (HgTe) and $m_h = 0.4\ m_o$, it follows from the condition (3.8) that in fields $H \lesssim 200$ kOe the energy of the acceptor level remains invariable. Thus the hole spectrum quantization and the $\mathcal{E}_A(H)$ dependence may be neglected for p-type GS in a wide and practically topical range of magnetic fields. In a field $H = H_c$ (Fig. 17) the bottom of the conduction band, as it moves upward, reaches the energy $\mathcal{E}_A$. The value of H_c is determined by the equality $\delta(H_c) = \mathcal{E}_A$ (for the determination of the gap $\delta(H)$ see Eqs. (2.35), (2.42), and (2.43)). At $H > H_c$ the acceptor level emerges in the field-induced gap and becomes sharp. For donor levels a sensible $\mathcal{E}_D(H)$ dependence should be expected already in fields of the order of several kilooersteds. However, because of the small effective mass m_e, n-type GS are heavily doped ($N_D^{1/3}\ a_{eB} > 1$) and the donor band is merged with the conduction band. Therefore the lower boundary of fields at which donor states separate from the conduction band depends on donor concentration N_D and is defined by the formula (3.11) for H_M.

Examine a compensated p-type GS with an impurity concentration $N_D < N_A < N_{AM}$ where N_{AM} is the acceptor concentration at which the Mott transition occurs (for HgCdTe $N_{AM} \simeq 10^{17} cm^{-3}$). Let the donor concentration N_D be larger than or equal to n_o, with n_o being the number of states in the conduction band below the acceptor level (see Sect. 3.6). Under these conditions at $T = 0$ K and $H = 0$ the Fermi level coincides with the acceptor level. As the magnetic field is increased, the bottom of the conduction band rises relative to the valence band, the number of states in the band with energies $\mathcal{E} \leq \mathcal{E}_A$ decreases and when $H > H_c$ (Fig. 17) all electrons occupy the acceptor states, i.e., ionize neutral acceptors. Since $N_D < N_A$, there is a sufficiently large number of unoccupied acceptor states and the Fermi energy is practically independent of magnetic field H.

We estimate the shift of the Fermi level at $H = H_c : \Delta\mathcal{E}_F = \mathcal{E}_F(H_c) - \mathcal{E}_F(0)$. The density of acceptor states is $\rho_A \simeq N_A/\Gamma$, where the width of the acceptor level, according to the Equation (3.2), is $\Gamma \sim (m_e/m_h)^{3/2}\mathcal{E}_A$. When all the n_o conduction band electrons pass over to the acceptors, $\Delta\mathcal{E}_F \simeq n_o/\rho_A \simeq (n_o/N_A)\Gamma$. For HgTe with $\mathcal{E}_A$ = 2 meV, $m_e = 0.03\ m_o$, $m_h = 0.4\ m_o$, the concentration $n_o = 2 \times 10^{15} cm^{-3}$ and even for the smallest $N_A = 1 \times 10^{16} cm^{-3}$ $\Delta\mathcal{E}_F \simeq 10^{-2}\ \mathcal{E}_A$. For GS HgCdTe the quantities n_o and Γ proportional to $m_e^{3/2}$ are smaller than the counterparts for HgTe and, consequently, $\Delta\mathcal{E}_F$ is smaller. This means the Fermi level in GS may be regarded to a high degree of confidence as fixed at the energy $\mathcal{E}_A$ with respect to the top of the valence band. Thus GS exhibit a very specific effect of the magnetic freeze-out of conduction band electrons into acceptors.

It is easy to understand that the Fermi level remains fixed in a magnetic field also in the case of a sufficiently wide acceptor band, even when its width Γ is comparable with $\mathcal{E}_A$. The Fermi level shift $\Delta\mathcal{E}_F$ due to the freeze-out of electrons into acceptors is negligibly small, i.e., $\mathcal{E}_F(H) \simeq \mathcal{E}_F(0)$. In this case the energy $\mathcal{E}_F(0)$ no longer coincides with the energy $\mathcal{E}_A$ for an isolated acceptor but is determined, as is usually the case with the degenerate electron gas, by the electron concentration n: $\mathcal{E}_F(0) \equiv \mathcal{E}(k_F)$, where the wavevector on the Fermi level is $k_F = (3\pi^2 n)^{1/3}$.

Inspection of Figure 17 shows that the value of field H_c at which the band electron concentration tends to zero may be used to determine $\mathcal{E}_A$ for small acceptor concentrations $N_A \ll N_{AM}$: $\mathcal{E}_A = \delta(H_c)$. When $N_A \gtrsim N_{AM}$ the freeze-out field depends only on electron concentration. For the nonquadratic dispersion relation (2.16) and the expression for the gap in a magnetic field (2.45) we find

$$H_c = \frac{2\hbar}{e}\left(3\pi^2 n\right)^{2/3}. \tag{3.22}$$

The magnetic field dependence of the concentration of electrons with the Fermi energy $\mathcal{E}_F(0) = \mathcal{E}_A$ under conditions of freeze-out into acceptors is given by the expression

$$n(H) = \frac{m_e^{1/2}}{\pi^2 L^2}\left[\mathcal{E}_F(0) - \delta(H)\right]^{1/2}. \tag{3.23}$$

Obviously, this dependence is of an altogether different character as compared to that in the intrinsic conductivity region where the electron gas is nondegenerate (2.47). The effect of the freeze-out of electrons into acceptors manifests itself in a very specific way in the magnetic field dependence of the longitudinal conductivity $\sigma_{zz} = \rho_{zz}^{-1}$. Measurements of $\sigma_{zz}(H)$ on GS $Hg_{0.86}Cd_{0.14}Te$ have shown that starting from some value of field the longitudinal conductivity σ_{zz} decreases linearly with H (Fig. 29) (Elliott et al., 1972). In a quantizing magnetic field the longitudinal conductivity $\sigma_{zz} = en\mu_e$ decreases with field not only as a result of the freeze-out of electrons according to the Equation (3.23) but also due to a change in their mobility μ_e. As a result, with the

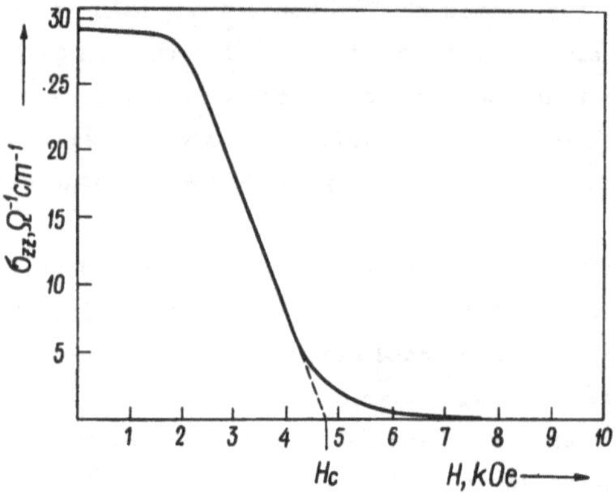

Fig.29. Magnetic field dependence of longitudinal conductivity for a $Hg_{0.86}Cd_{0.14}Te$ sample

Fermi level being fixed in a magnetic field, $\mathcal{E}_F(H) = \mathcal{E}_F(0)$, the longitudinal conductivity $\sigma_{zz} \sim [\mathcal{E}_F(0) - \delta(H)|$. Since the gap $\delta(H)$ increases linearly with field, the conductivity should really decrease in a linear fashion.

By extrapolating the linear part of the curve in Figure 29 to the abscissa axis, it is possible to determine the freeze-out field H_c. For all the $Hg_{0.86}Cd_{0.14}Te$ samples that have been investigated, the values of H_c thus found agree well with those calculated from the known electron concentration with the help of the Equation (3.22). The observed linear law of decreasing $\sigma_{zz}(H)$ and the experimentally confirmed relation between H_c and n give another testimony to the Fermi energy $\mathcal{E}_F$ being independent of field H in the HgCdTe samples explored.

In magnetic fields $H>H_c$ the degeneracy of the electron gas is removed even at very low temperatures (the Fermi level descends below the bottom of the conduction band) and the electron concentration decreases exponentially with increasing field due to the freeze-out of electrons. When $\mathcal{E}_F(H) = \mathcal{E}_F(0)$ and with neglect of the comparatively weak H dependence of the preexponential factors, the expression for the longitudinal magnetoresistance $\rho_{zz}(H)$ has the form

$$\rho_{zz}(H) \sim \exp\left[\frac{\delta(H)}{k_B T}\right] . \tag{3.24}$$

The origin of the exponential portion of decreasing σ_{zz} in Figure 29 coincides approximetely with the deviation of the $\sigma_{zz}(H)$ curve from the dotted straight line which corresponds to the strong-degeneracy condition.

It is seen from a comparison of the formulas (3.24) and (2.48) that to find the gap δ from the exponential $\rho_{zz}(H)$ dependence one has to know what type of conductivity (intrinsic or extrinsic) occurs in the temperature interval studied. If this is not taken into account, one may err doubly in estimates of the δ value. The freeze-out of electrons into acceptors at $H>H_c$ was observed for GS $Hg_{0.9}Cd_{0.1}Te$. In GS HgCdTe the electron effective masses and the gap $|\mathcal{E}_g|$ values are smaller than the counterparts in HgTe, and the magnetic-field-induced gap is larger. Therefore the preexponential factor does not practically distort the exponential dependence $\rho_{zz}(H)$.

Figure 30 presents $\rho_{zz}(H)$ dependences for a $Hg_{0.9}Cd_{0.1}Te$ crystal (Gorodilov et al., 1980). Over the 20-100 K temperature range all the curves exhibit segments of activational growtH of the magnetoresistance according to the law (3.24). From the slope

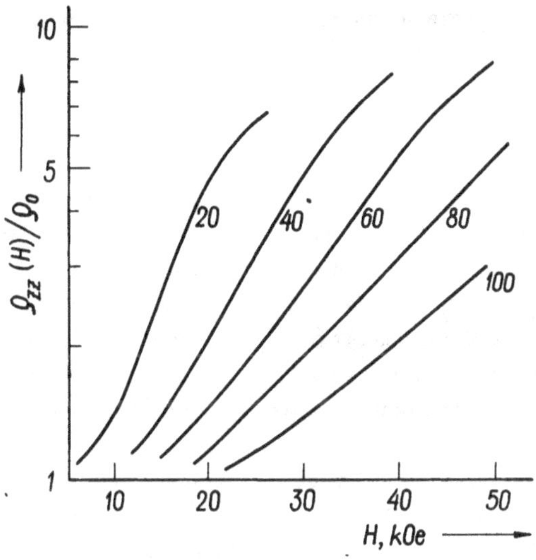

Fig.30. $\rho_{zz}(H)$ relation for GS $Hg_{0.9}Cd_{0.1}Te$. The numerals at the curves indicate temperature, K; ρ_0 is the resistivity at $H = 0$.

of the rectilinear segments of the $\ln \rho_{zz}(H)$ curves, one can determine the gap as a function of magnetic field. It has turned out that the gap δ varies in proportion to field H. The $\delta(H)$ relation is well described by the formula (2.43) (which holds for small $|\mathcal{E}_g| \lessapprox 100$ meV) if one chooses the following values of the parameters involved in the expression for $\bar{\gamma}$ (2.40): $\mathcal{E}_p = 18$ eV, $\mathcal{E}_g(T{=}0) = -110$ meV, and $(d\mathcal{E}_g/dT) = 0.4$ meV deg^{-1}. The conformance to the exponential law (3.24) in the temperature interval specified indicates that the Fermi energy is independent of magnetic field. The region of intrinsic conductivity in the samples with $N_A \approx 10^{17} cm^{-3}$ that have been studied, for which the formula (2.48) specifying the $\rho_{zz}(H)$ dependence holds, begins at room temperature.

The electron spectrum readjustment in an intrinsic gapless semiconductor under the effect of hydrostatic pressure was discussed in Sect. 2.3.7. If GS contain impurities, the pressure dependences of the Fermi energy, electron concentration, and a number of transport phenomena, specifically conductivity, are different for p-type and n-type crystals. For GS $Hg_{1-x}Cd_xTe$ with $0.1 \le x \le 0.16$, which are of greatest interest from the point of view of studying the GS-semiconductor transition, since the gap is small ($|\mathcal{E}_g| \lessapprox 100$ meV), the electron dispersion relation is nonquadratic and is described with good accuracy by the Equation (2.15). Allowing for the dependence of the gap $\mathcal{E}_g$ on pressure $\mathcal{P}$ (2.49), the relation between the electron concentration and the Fermi energy may be represented as

$$(3\pi^2 n)^{2/3} = \frac{3\mathcal{E}_F}{2P^2}(\mathcal{E}_F - \mathcal{E}_g(0) - \beta\mathcal{P}) \quad . \tag{3.25}$$

For p-type samples ($N_A > N_D$) the Fermi level is located in the region of a large density of acceptor states, $\rho_A(\mathcal{E}_F)$, which considerably exceeds the density of states in the conduction band, $\rho_c(\mathcal{E}_F)$. Therefore the Fermi level turns out to be independent of pressure,

$$\mathcal{E}_F(\mathcal{P}) = \mathcal{E}_F(\mathcal{P} = 0) = \text{const} \quad , \tag{3.26}$$

for the same reasons as those considered above in this section when discussing the stabilization of the Fermi level in a magnetic field.

Subject to the condition (3.26), it clearly follows from the expression (3.25) that the electron concentration in p-type GS decreases with increase in pressure since $\mathcal{E}_g$

(but not the modulus $|\mathcal{E}_g|$) increases with $\mathcal{P}$. At some pressure $\mathcal{P} = \mathcal{P}_n$, which is determined by the Equation .

$$\mathcal{E}_F(\mathcal{P}=0) = \mathcal{E}_g(\mathcal{P}_n) = \mathcal{E}_g(0) + \beta\mathcal{P}_n \quad , \tag{3.27}$$

the electron concentration tends to zero (at T=0 K). The conductivity and the Hall effect in this case are determined by low-mobility charge carriers - valence or acceptor band holes. The pressure $\mathcal{P}_n$ at which the electron concentration goes to zero is normally not in large excess over the pressure $\mathcal{P}_c$ (see Eq.(2.50)) at which the gap $\mathcal{E}_g$ vanishes. This is so because, as a rule, the Fermi energy $\mathcal{E}_F$ ($\mathcal{P}$=0) $\leq$ 10 meV[23]. For temperatures T>0 K and at $\mathcal{P} > \mathcal{P}_n$ the Fermi level enters into the pressure-induced gap between the s- and p-bands, and the electron concentration remains a finite although exponentially small quantity.

The linear $n^{2/3}$ dependence of $\mathcal{P}$ was established for a number of gapless semiconductors $Hg_{1-x}Cd_xTe$ with $0.10 \leq x \leq 0.15$ (Elliott et al., 1972; Brandt et al., 1976). The hole concentration $p = N_A - N_D$ is normally determined from the Hall coefficient $R = 1/ep$ at 77 K when all the acceptors are ionized; the electron concentration n is found from the period of Shubnikov - de Hass oscillations. If the parameters β and P^2 are known, the Fermi energy $\mathcal{E}_F(\mathcal{P}=0)$ can be determined from the slopé of the $n^{2/3}(\mathcal{P})$ curves The intercept of the straight lines $n^{2/3}(\mathcal{P})$ with the abscissa axis corresponds to the pressure $\mathcal{P}_n$, and using the Equation (3.27) one can find the gap $\mathcal{E}_g(0)$ at atmospheric pressure.

Hall effect, conductivity, and magnetoresistance measurements on p-type GS HgCdTe have permitted the conclusion that in weakly doped (and moderately compensated) samples with $N_A - N_D \approx 10^{16} cm^{-3} \ll N_M$ for which the density of states has the shape such as that shown in Figure 27b) the Fermi level coincides with the acceptor level. This allows $\mathcal{E}_F$ to be determined from the $\mathcal{E}_A$ value found. In heavily doped and highly compensated samples with $N_A - N_D > 5 \times 10^{17} cm^{-3} > N_M$ for which the density of states is portrayed in Figure 27d) the Fermi level lies deep within the density of states tail of the valence band, $\mathcal{E}_F \simeq \gamma$.

In n-type GS with $N_D \gg N_A$ the Fermi level emerges in the region of energies where the density of states in the conduction band, $\rho_c(\mathcal{E}_F)$, largely exceeds the density of acceptor states $\rho_A(\mathcal{E}_F)$. Since not only all donors but also all acceptors are ionized in such a situation, the electron concentration $n = N_D - N_A$ does not depend on pressure:

$$n(\mathcal{P}) = \text{const.} \tag{3.28}$$

According to the Equation (3.25) and (3.28), the Fermi energy $\mathcal{E}_F$ in n-type GS should first increase with pressure, reaching a maximum value at $\mathcal{P} = \mathcal{P}_c$ ($\mathcal{E}_g = 0$); upon transformation of a zero-gap semiconductor to a finite-gap semiconductor, the quantity $\mathcal{E}_F$ should start decreasing. It is this variation of $\mathcal{E}_F$ with $\mathcal{P}$ that has been established for $Hg_{1-x}Cd_xTe$ crystals with $x \simeq 0.12$ to 0.13 (Stankiewiecz et al., 1971; Carvalho et al., 1982).

23) Such values occur in highly compensated GS on account of a large magnitude of the random potential γ.

On the whole, investigations of GS subject to hydrostatic pressure have shown that there is a considerable overlap, such as that depicted in Figure 27d), of the conduction band and the impurity-perturbed valence band. Only at energies $\mathcal{E} > 10$ to 20 meV is the conduction band free from acceptor states.

3.9. Low-Temperature Peculiarities of Conductivity

3.9.1. Conductivity Minimum

GS of the p-type ($N_A > N_D$) repeatedly exhibited at low temperatures the following two peculiarities of the temperature dependence of conductivity:a minimum of $\sigma(T)$ in the 5-11 K range and a plateau in the 20-40 K range. These features were treated in the literature in different ways. Rather than go into the particulars of the discussion we outline the viewpoint which is held to be most correct (Arapov et al., 1982; Brandt et al., 1983).

We start with the $\sigma(T)$ minimum. There is no clear-cut correlation between the temperature T_{min}, at which the conductivity is minimal, and the composition of the $Hg_{1-x}Cd_xTe$ solid solution: For crystals with a Cd content ranging from x = 0 (HgTe) to $x \approx 0.155$, the $\sigma(T)$ minimum occurs in the same temperature region. The same is true of GS HgMnTe on which this minimum was studied thoroughly. On the other hand, both the position (T_{min}) and the depth of the minimum may vary somewhat for different samples of the same composition. Thus for $Hg_{1-x}Cd_x$ Te crystals with $0.05 \leq x \leq 0.16$, T_{min} varies from 5 to 11 and its depth $\sigma(1.7\ K)/\sigma(T_{min})$ ranges between 1.2 and 3.3. The deepest minima occur in samples of GS $Hg_{1-x}Cd_xTe$ with small $|\mathcal{E}_g|$ ($x \approx 0.16$). Figure 31 presents the $\sigma(T)$ dependence for a p-type $Hg_{1-x}Cd_xTe$ sample with x = 0.15.

Deeper $\sigma(T)$ minima have been revealed in p-type GS $Hg_{1-x}Mn_xTe$ with $0.036 \leq x \leq 0.075$. With increase in temperature in the region T = 0.05-0.2 to T = 4-8 K, the specific resistivity increases by a factor of 1.5-7. The magnetic field dependence of the Hall coefficient indicates that in the HgMnTe crystals that have been investigated the

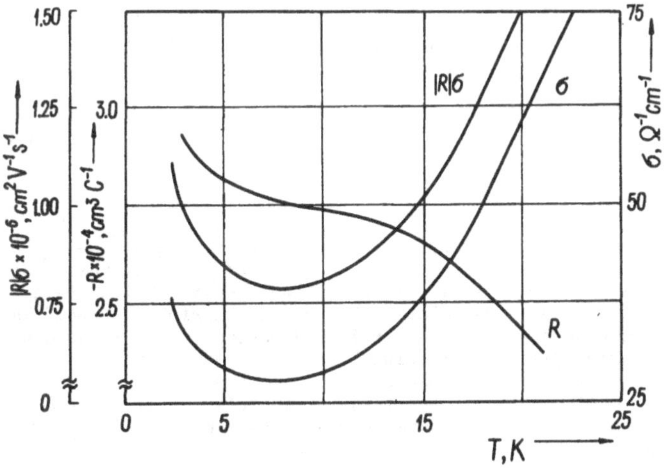

Fig.31. Temperature dependence of mobility σ, Hall coefficient R, and Hall mobility $|R|\sigma$ for $Hg_{0.85}Cd_{0.15}Te$ crystal

conduction occurs by two types of charge carriers with largely differing mobilities[24]: electrons with mobilities 10^5-10^6 $cm^2V^{-1}s^{-1}$ and acceptor band holes with mobilities 10^2-10^3 $cm^2V^{-1}s^{-1}$. By the character of the temperature dependences of the specific resistivity ρ and Hall coefficient R, all the HgMnTe studied may be divided into two groups.

One group involves samples in which ρ increases by a factor of 2-5 when the temperature is increased approximately from 0.1 to 5 K, and $|R|$ in weak fields (20-50 Oe) in the same temperature interval decreases abruptly (by a factor of 5-70). In samples of the other group the Hall coefficient does not vary with increase in temperature from 0.1 to 5 K and the resistivity increases by a factor of 5-7. Typical dependences for both groups of samples are presented in Figure 32. An analysis has shown that in the first group of samples the partial conductivities of conduction band electrons, $\sigma_e = en\mu_e$, and heavy holes, $\sigma_h = en\mu_h$, are comparable in magnitude, and in the second group $\sigma_e \gg \sigma_h$.

Investigations of Shubnikov - de Haas oscillations have shown that for all the samples that have been studied the oscillation period in the interval $0.1 \leq T \leq 5$ K is independent of temperature. Consequently, the electron concentration is constant, too. It is therefore natural to assume the observed growth of $\rho(T)$ and the decrease in $|R(T)|$ to be due to a change in conduction band electron mobility μ_e. When two types of charge carriers participate in conduction, the conductivity and the Hall effect in weak fields $H \to 0$ are determined by the expression

$$\sigma = e(n\mu_e + p\mu_h) , \tag{3.29}$$

$$R = \frac{e(n\mu_e^2 - p\mu_h^2)}{\sigma^2} . \tag{3.30}$$

The formula (3.30) is written for charge carriers of different sign. For the samples that have been studied, the electron mobility largely exceeds the heavy-hole mobility, so that $n\mu_e^2 \gg p\mu_h^2$.

It follows from Equations (3.29) and (3.30) that the unusual behavior of $\sigma(T)$ and $R(T)$ at low temperatures is due chiefly to a change in band electron mobility μ_e. According to the Equation (3.30), the mobility μ_e is expressed in terms of the observables R, σ, and n (n is determined from the Shubnikov - de Haas oscillation period) as

$$\mu_e(T) = \left[\frac{|R(T)|\ \sigma^2(T)}{en}\right]^{1/2} . \tag{3.31}$$

24) If one type of charge carriers, say electrons, participate in the conduction, the Hall coefficient R=1/en in the absence of a freeze-out of electrons is independent of magnetic field. But if along with high-mobility electrons also "heavy" holes (electrons) contribute to the conduction, the R(H) relation has a characteristic shape. In weak fields where only light electrons are essential, R is independent of H. In stronger fields, as a result of the Larmor twisting of "light" electrons and owing to a decrease in their mobility,the contribution of these electrons to the Hall e.m.f. decreases and $|R|$ diminishes. In sufficiently strong fields it is only "heavy" carriers that contribute to the Hall effect, and, again, R is independent of H. Thus the R(H) curve consists of two horizontal segments connected by a smooth sloping curve. When the R vs. H curve has such a shape one may be sure that at least two charge carrier types participate in the conduction.

The $\mu_e(T)$ dependences, calculated using the Equation (3.31), are presented in Figure 32. It is seen that with decreasing temperature the electron mobility increases abruptly, and this is the major cause for the anomalous temperature trend of σ and R. An explicit correlation exists between the variation of the mobility and the σ_h/σ_e ratio. The larger the contribution made to charge transport by the heavy holes σ_h which presumably move over the acceptor band, the more strongly the mobility varies with temperature. Also, it might be well to point out that the low-temperature conductivity minimum occurs only in GS. When the latter are converted to finite-gap semiconductors (such experiments have been performed on GS $Hg_{0.9}Cd_{0.1}Te$ under hydrostatic pressure) the $\sigma(T)$ minimum vanishes.

A natural explanation of the cause for the formation of a $\sigma(T)$ minimum suggests itself. At low temperatures, $T<T_{min}$ the electron mobility decreases with increasing temperature, and at $T>T_{min}$ an increase in electron concentration n occurs which results from the excitation of electrons from the valence band into the conduction band. The problem now reduces to ascertaining why μ_e decreases anomalously with increase in T.

Evidently, one has to assume that in the region of such low temperatures, where electrons are scattered primarily by charged centers, the number of the latter varies in some way with temperature. It seems natural that with increasing temperature in the region $T<T_{min}$ the number of electrons excited from the valence band onto acceptors grows and, hence, the number of ionized acceptors, N_A^- ; increases. However, this mechanism of N_A^- growth is ineffective since the mean thermal energy k_BT is much less than the acceptor level energy $\mathcal{E}_A$. Therefore one has to look for another reason for the increase in the number of scattering centers.

An idea was advanced that donor-acceptor complexes in gapless semiconductors play a specific role. Since all donors in GS are ionized, part of the electrons that left the donors fill the conduction band, and the rest of the electrons are captured by acceptors and ionize these. In contrast to normal semiconductors where charged acceptors arise only as a result of the thermal excitation of electrons from the valence band,

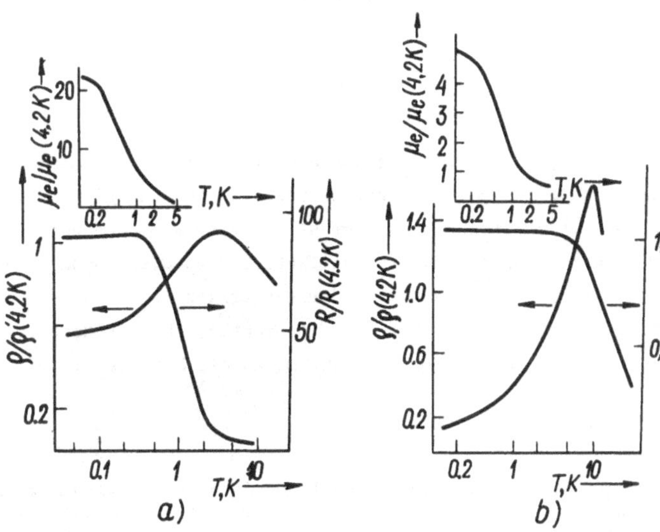

Fig.32. Temperature dependence of specific resistivity ρ, Hall coefficient R, and electron mobility μ_e (calculated according to formula (3.31) for GS $Hg_{1-x}Mn_xTe$: a) x=0.036, b) x=0.065

charged acceptors in GS at sufficient concentrations N_D and N_A may arise at the lowest temperatures.

A minimum of electrostatic energy will be assured when negatively charged acceptors settle down near positively charged donors. Thus a donor-acceptor pair is an electric dipole. In the HgMnTe samples that have been studied, the concentration N_A is much larger than N_D, and the dimension of the dipole (the mean distance between the donor and the acceptor) $d \approx N_A^{-1/3}$. The band electron wavelength is $\lambda = 1/k_F \approx 1.8 \times 10^{-5}$ cm (with $\hbar k_F$ being the momentum on the Fermi level), and for concentrations $3 \times 10^{16} \leq N_A \leq 1 \times 10^{17}$ cm^{-3} the dipole dimension $d \approx (0.1-0.3) \times 10^{-5}$ cm, i.e., $\lambda > d$. Under such conditions the electron does not "feel" each dipole-forming ion taken separately but is scattered on the dipole as a whole. Since the dipole is neutral, the probability of scattering on it is smaller than the charged-center scattering probability. A substantial contribution to scattering at ultralow temperatures is therefore made by positively charged donors that are not bound into pairs ($N_D > N_A^-$ since part of the electrons have withdrawn from the donors into the conduction band).

As the temperature is raised, quasibound electrons hop from charged acceptors on to neutral ones in the vicinity of which there are no donors (let it be recalled that the total number of acceptors $N_A >> N_A^-$). Neutral dipoles collapse, the number of separate charged centers serving as sources of conduction band electron scattering grows, and the charged-center scattering probability increases[25]. This is what is responsible for the decrease in electron mobility with increasing temperature.

It is obvious that the more donor-acceptor complexes collapse to form $N_D+N_A^-$ separate charged centers, the stronger the decrease in μ_e will be. It is therefore in samples whose acceptor band conductivity σ_h is comparable with the band electron conductivity σ_e that the decrease in mobility with increasing temperature is much more dramatic than that in samples with $\sigma_e >> \sigma_h$. In the first case, i.e., when $\sigma_h \approx \sigma_e$, there are more electrons in the acceptor band and, accordingly, a larger number of ionized acceptors involved in complexes which form a large number of scattering charged centers when the temperature is increased.

3.9.2. A Plateau on the σ(T) Curves

Now we turn to the second peculiarity of GS at low temperatures, viz., the plateau (step) on the σ(T) curves at 20-40 K (Fig. 33). In contradistinction to the σ(T) minimum at $T \approx 5-11$ K, which disappears at the transition from a zero-gap semiconducductor to a finite-gap semiconductor, the plateau at 20-40 K manifests itself also when the gap between the conduction and valence bands opens up under the effect of a magnetic field or hydrostatic pressure. In accounting for the anomaly of σ(T), the authors of pioneer investigations proceeded from the assumption that the main role here is played

25) At low temperatures donor-acceptor complexes form also in semiconductors with a gap. But, in contrast to GS, semiconductors with $N_a > N_D$ at low temperatures ($T \to 0$) do not have conduction band electrons that serve in GS as an indicator of a destruction of dipoles as the temperature is elevated.

by conduction electrons and the $\sigma(T)$ anomaly is due to features peculiar to the temperature dependence of either their concentration or their mobility (Finck et al., 1972; Dietl et al., 1978). For example, as a result of electron resonance scattering on acceptors forming a hypothetical level (of energy 9.5 meV for HgTe), not detected in magnetooptical measurements, the electron mobility should display a minimum in the 20-40 K range. The anomaly of $\sigma(T)$ can be explained consistently if one allows for the unusual behavior of the Hall coefficient R(T) in the same temperature interval.

Figure 33 presents the temperature dependences of the longitudinal conductivity σ_{zz} and Hall coefficient R in various magnetic fields for a $Hg_{0.845}Cd_{0.155}Te$ sample (Arapov et al., 1982). As can be seen from the figure, the increase of $\sigma(T)$ (curve 1) slows down in the 20-40 K range and the $\sigma(T)$ curve exhibits a peculiarity as a step. In a magnetic field the step on the $\sigma_{zz}(T)$ curves first smoothens (curves 2,3), and in sufficiently strong fields $H \gtrsim 10$ kOe it becomes clearly pronounced again. Over the same temperature interval the R(T) curves exhibit a crevasse which deepens with increasing field and then succeeds to a region of positive R values (dotted curves). The Hall coefficient as a function of temperature undergoes double reversal of sign.

It is a simple matter to see that neither the behavior of the Hall effect nor that of the longitudinal conductivity can be explained on the assumption that only band electrons participate in charge transport. Let us assume that along with light conduction band electrons the contribution to charge transport phenomena is made also by heavy charge carriers and let us try to ascertain whether the temperature anomaly of $\sigma(T)$ can be explained on the basis of such an assumption. Analysis of the R(H) dependence permits one to

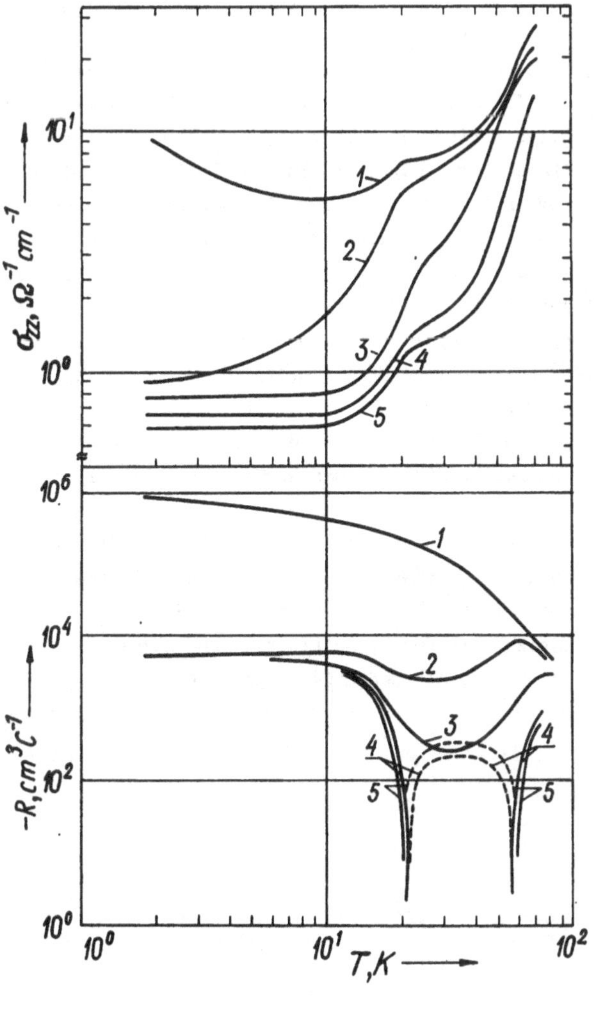

Fig.33. $\sigma_{zz}(T)$ and R(T) relations in different magnetic fields for a $Hg_{0.845}Cd_{0.155}Te$ crystal. H, kOe: (1) 0 for σ_{zz} and 0.01 for R, (2) 1.5, (3) 6, (4) 15, (5) 22. The dashed lines show segments of the R vs.T curves where R>0.

judge what kind of charge carriers participate in conduction at a given temperature. If the magnetic fields are so weak that the concentrations and mobilities of light and heavy charge carriers do not change in a magnetic field, the R(H) dependence may be written in the form

$$\frac{R(0)}{R(0) - R(H)} = a + \frac{s}{H^2} . \tag{3.32}$$

The formula (3.32) is valid for magnetic fields in which the opening gap $\delta(H)$ is less than the Fermi energy $\mathcal{E}_F$. Otherwise the Fermi level enters into the gap and the band electrons start being frozen out into acceptors. On plotting the dependence (3.32) one can determine the parameters a and s from the R(H) values found experimentally and then calculate the concentrations n_1, n_2 and mobilities μ_1, μ_2 of light and heavy charge carriers using the measured quantities σ and R(0). The value of a found at $T \lessapprox 15$ K is larger than unity. Since $R(0) < 0$, $R(H) < 0$, too (see Equ.(3.32)). Consequently, both kinds of charge carriers are electrons, light and heavy. The calculated value of light-electron concentration n_1 coincides with the quantity n_e determined from the Shubnikov - de-Haas oscillation period. For $45 \text{ K} \leq T \leqq 60$ K $a < 1$, and therefore, in addition to light electrons, holes participate in conduction.

In the region of intermediate temperatures $15 \text{ K} < T < 45$ K the R versus H curves cannot be described by the Equation (3.32). The idea suggests itself that this comes from one more species of charge carriers participating in conduction, viz., heavy electrons. But what sort of heavy electrons are these charge carriers? In the samples studied, $n_1 < N_D < N_A$ and so the Fermi level is located in the acceptor band. For magnetic fields in which $\delta(H) > \mathcal{E}_F$, the acceptor states enter into the gap that has opened up, the light electrons are frozen out into acceptors, and thus the heavy electrons governing, as has already been mentioned, the R(H) behavior at $T < 15$ K are nothing else but acceptor band charge carriers (the $T < 15$ K concentration of valence band holes arising from the excitation of valence band electrons into acceptors is negligibly small)[26]. This conclusion is supported also by the fact that when $T > 45$ K the impurity band is increasingly filled and its contribution to the conductivity decreases to zero. At such temperatures, as an analysis of the Hall effect (3. 32) has shown, heavy electrons do not participate in conduction.

Now we try to ascertain the cause for the occurrence of a step on the σ_{zz} versus T curves. Let the magnetic field be strong enough that $\delta(H) \gg \mathcal{E}_F$, $k_B T$ and all the band electrons be frozen out into the acceptor states located in the gap $\delta(H)$. Inasmuch as we assume heavy electrons to be electrons captured by acceptors, the heavy-electron concentration is $n_2 = N_A^- = N_D + p$. The longitudinal conductivity then is

$$\sigma_{zz} = e(N_A^- \mu_2 + p\mu_p) . \tag{3.33}$$

26) We have no reasons for ruling out the possibility that heavy electrons are charge carriers of the surface inversion layer of the sample. The problem calls for further exploration.

At temperatures $k_BT \ll \mathcal{E}_A$ ($\mathcal{E}_A$ = 6 meV corresponds to ≈ 70 K) the hole concentration p is negligible and it follows from the Equation (3.33) that

$$\sigma_{zz} = eN_D\mu_2. \tag{3.34}$$

According to the Equation (3.33) the $\sigma_{zz}(T)$ behavior is determined primarily by the temperature dependence of N_A^- and p. At low temperatures ($k_BT \ll \mathcal{E}_A$) the conductivity σ_{zz} is, according to the Equation (3.34), independent of temperature. As the temperature is raised the hole concentration p starts growing and σ_{zz} increases. The slowdown in the growth of $\sigma_{zz}(T)$ at $T \gtrsim 20$ K is associated with a transition to a region where acceptors are saturated with electrons. In the case of complete saturation all acceptors are ionized, i.e., all acceptor states are populated with electrons. When T > 40 K light valence band electrons appear in the conduction band, leading to an increase in σ_{zz} (Fig.33).

Calculations of the $N_A^-(T)$ and p(T) dependences for the case when the magnetic field is absent show that the σ versus T curve at T < 40 K should have a step similar to that on the σ_{zz} versus T curve in the presence of a magnetic field. This permits the conclusion that the σ(T) peculiarity at H = 0 is of the same nature as that at H ≠ 0, i.e., is due to heavy charge carriers.

It remains to be explained why the step on the $\sigma_{zz}(T)$ curves in intermediate fields is less pronounced than that in high fields or in a zero field. In strong magnetic fields ($\delta(H) \gg k_BT$, $\mathcal{E}_A$) the situation is the same as that in a semiconductor with a gap $\mathcal{E}_g \gg \mathcal{E}_A$, when the acceptor level is far from the conduction band: At temepratures $k_BT \ll \delta(H)$ the electrons are practically not excited into the conduction band, and n_1=0. The growth of σ_{zz} with temperature at T < 20 K is determined by the exponential increase in the number of acceptor band electrons, n_2, and in the number of valence band holes, p. With the temperature elevated above 20 K, the increase of $n_2(T)$ and p(T) slows down due to the saturation of acceptors with electrons and, therefore, the growth of $\sigma_{zz}(T)$ slows down, too.

In the absence of a magnetic field, when the semiconductor is gapless, light electrons obey the Fermi-Dirac statistics and their concentration n_1 varies very weakly with temperature, whereas the electron concentration n_2 of the acceptor band increases exponentially until the latter becomes filled up, as a result of which a step arises on the σ versus T curve. In intermediate magnetic fields such that $\sigma(H) \approx \mathcal{E}_A$, the acceptor states are sufficiently close to the conduction band and the rates at which the electron and hole concentration increase with temperature at T < 20 K are comparable in magnitude: $\Delta n_1/\Delta T \approx \Delta n_2/\Delta T \approx \Delta p/\Delta T$. In this case the saturation of acceptors with electrons as the temperature is raised does not give rise to a well-defined step on the σ_{zz} versus T curve, i.e., the step smoothens because of the continuous increase of the concentrations n_1 and p (although n_2 ceases to depend on T).

Thus the anomalous behavior of σ(T) as well as R(T) in GS HgCdTe in the region of intermediate temperatures 20-40 K is due to the contribution which heavy charge carriers - impurity band electrons and valence band holes - make to transport processes. The most important argument to support this conclusion is the presence of a quite similar conductivity anomaly $\sigma_{zz}(T)$ in a magnetic field. In fields for which the gap $\delta(H) \gg k_BT$, there are few or no electrons in the conduction band - they are frozen out, and the peculiarities in the variation of their concentration, $n_1(T)$, or mobility, $\mu_1(T)$, cannot therefore be responsible for the $\sigma_{zz}(T)$ anomaly.

The question arises whether the peculiarity of the conductivity at 20–40 K appears at any acceptor and donor concentration. An analysis of experimental data for GS HgTe and HgCdTe of the p-type shows that this $\sigma(T)$ peculiarity is most clearly marked in the purest samples with the smallest acceptor and donor concentrations ($N_A \lesssim 10^{16} cm^{-3}$ and $N_D \lesssim 10^{15} cm^{-3}$). When $N_A > 10^{17} cm^{-3}$ the step disappears. The $\sigma(T)$ peculiarity described above may be thought to occur only when there are not very many acceptors and the compensation ($N_D << N_A$) is small, so that the acceptor band is separated from the valence band. The activational dependence of the heavy-electron and hole concentrations, n_2 and p, is possible here. At large N_A and when the compensation is considerable ($N_D \approx N_A$) the acceptor states merge with the valence band and a distorted valence-band-edge potential relief forms.

The following important circumstance must be noted. The $\sigma(T)$ feature at 20–40 K can be accounted for on the assumption that heavy electrons play a substantial role in charge transport. We assume in this case that they are located in the acceptor band. In other p-type semiconductors the conduction via the acceptor band occurs by holes, i.e., the Hall effect is positive. As a matter of fact, theory admits the possibility of the Hall effect being of negative sign in the case of conduction via acceptor states, but this conclusion relates to the case of hopping conductivity, when the acceptors are few and the acceptor wavefunctions overlap weakly (no impurity band arises). In this situation the mobility of charge carriers is very low ($\mu \approx (1-10^2)\ cm^2V^{-1}s^{-1}$) and depends exponentially on temperature. Under the conditions discussed, heavy electrons possess comparatively high mobilities $\mu_2 \approx (10^3-10^4)\ cm^2V^{-1}s^{-1}$, which indicates that the conductivity is most likely to be metallic in character. In this context it remains to be ascertained whether heavy electrons are acceptor band charge carriers.

A generally prevalent standpoint is that the double sign reversal of the Hall effect arises from the contribution of n-type conductivity surface layer electrons in p-type crystals (Scott and Hager, 1971). In a number of experiments, for example on p-$Hg_{0.8}Cd_{0.2}Te$ crystals, this assumption has undoubtedly been corroborated. At the same time there is reliable supportive evidence that heavy electrons are in the bulk of the crystal rather than in the surface layer. This has been established, for example, in measurements of GS $Hg_{1-x}Mg_xTe$ (Gluzman et al., 1981). Thus the problem calls for further exploration.

We conclude this section by briefly discussing the problem as to the magnitudes of hole mobilities in GS HgCdTe at low temperatures. For band electrons, the theory based on the concept of the relaxation time τ (or the free path length $l = v\tau$, with v being the mean electron velocity) leads to values of mobility, $\mu_e \approx 10^5-10^6 cm^2V^{-1}s^{-1}$, that are consistent with experimental data for noncompensated samples. The hole mobilities μ_h found experimentally for GS HgCdTe with impurity concentrations $N_A-N_D \gtrsim 1 \times 10^{17} cm^{-3}$ at helium temperatures are two or three orders of magnitude lower than the electron mobilities, i.e., are equal to 10^2-10^3 and sometimes even $\approx 10^2\ cm^2V^{-1}s^{-1}$. At low temperatures $T \lesssim 10$ K both electrons and holes are scattered chiefly by ionized impurity centers. According to theory, the mobility under such conditions is $\mu \sim m^{-1/2}$. If one takes into consideration the fact that the effective mass ratio $(m_h/m_e) \approx 10$ (for HgTe, for example, $m_h/m_e \doteq 13$), then such large ratios $(\mu_e/\mu_h) \approx 10^2-10^4$ appear to be surprising. The cause for the relatively low values of mobility μ_h lies very likely in the fact that the holes, which manifest themselves in kinetic effects at low temperatures, are in the valence band density of states tail. Therefore the dynamic properties of holes may be characterized by the band effective mass m_h, and the energy of teir interaction with scattering centers is large compared to the kinetic energy. Put another way they may not be treated as free (band) charge carriers and their motion may not be described with the help of the Boltzmann kinetic equation in the relaxation time approximation.

What is the lowest hole mobility allowed by the free path length concept? The usual expression for the hole mobility may be transcribed in terms of the relaxation time $\mu_h = e\tau/m_h$ into

$$\mu_h = \frac{e}{2\pi h} \, l\lambda , \qquad (3.35)$$

where $\lambda = 2\pi\hbar/m_h v$ is the de Broglie wavelength. The holes may evidently be viewed as delocalized, i.e., freely moving, if the wavelength is not less than the mean interimpurity distance: $\lambda \geq N_A^{-1/3}$. We assume here that only acceptors are impurities. In its very sense, the free path length l of delocalized holes may not be less than the interimpurity distance $N_A^{-1/3}$, either. Thus the minimum mobility of delocalized holes (which is said to be metallic or diffusive) should, according to the Equation (3.35), be

$$\mu_{min} \simeq \frac{e}{2\pi\hbar} N_A^{-2/3} \quad . \qquad (3.36)$$

The quantity μ_{min} does not depend on either the effective mass or other characteristics of the electron spectrum but is determined by the impurity concentration only. For $N_A \simeq N_M \approx 10^{17} cm^{-3}$ the Equation (3.36) yields $\mu_{min} \simeq 10^3 cm^2V^{-1}s^{-1}$. This exceeds somewhat the typical μ_h values found experimentally by different authors at 4.2 K for GS HgCdTe with approximately the same impurity content. Hence it follows that holes are localized on acceptors and the conductivity is of a hopping nature: As the temperature is increased holes hop from one acceptor to another. The conductivity in this case depends exponentially on temperature, by contrast with the metallic conductivity.

It should at the same time be borne in mind that the formula (3.36) holds to within a numerical factor of the order of unity and when there is no compensation ($N_D = 0$). In most of the cases the samples that have been investigated are compensated, to a larger or lesser degree. As a rule, the degree of compensation N_D/N_A is unknown and there are only estimates of $N_A - N_D$ from the value of the hole concentration in the region of temperatures (normally $\gtrsim$ 77 K) where all acceptors are ionized.

For GS HgCdTe it has been found experimentally that at acceptor concentrations $N_A \simeq N_M$ the impurity conductivity due to holes is of a metallic character if $\mu_h \gtrsim$ 300 $cm^2V^{-1}s^{-1}$. With $\mu_h \lesssim$ 100 $cm^2V^{-1}s^{-1}$ the hole conductivity σ_h and, accordingly, the mobility $\mu_h = R\sigma_h$ increase rather rapidly with temperature in the region T < 20–30 K. The Fermi level in this case is located in the valence band density of states tail.

3.10. Gapless Semiconductors Containing Magnetic Ions

Solid solutions of mercury chalcogenides (for example HgTe) and transition or rare-earth elements chalcogenides (such as MnTe and EuTe) form dilute magnetic semiconductors, or, as they are often called, semimagnetic semiconductors (SMS). Mercury chalcogenides are zero-gap semiconductors. Transition or rare-earth metal chalcogenides are magnetic semiconductors with a wide energy gap, i.e., are essentially dielectrics. By varying the composition of these solid solutions, it is possible to reconstruct the electron spectrum from gapless to semiconductive.

The interest in semimagnetic semiconductors (for concreteness, we will be concerned with the best studied SMS, viz., HgMnTe with a small Mn content) comes from the specificity of their properties (see the reviews by Furdyna, 1982; Mycielski, 1982, Brandt and Moshcalkov, 1984; Lyapilin and Tsidilkovski, 1985). On the one hand, in contrast to the conventional magnetic semiconductors, the band structure of the SMS (HgTe) host crystal has been reliably determined. On the other hand, the high mobility of band electrons in SMS, which exceeds the counterpart in normal magnetic semiconductors by many orders of magnitude, permits one to employ practically the whole arsenal of modern methods for investigating physical properties of solids. Specifically, an opportunity offers of determining very reliable the exchange interaction parameters, viz., the. exchange integrals.

In their properties, HgMnTe solid solutions resemble in many ways HgCdTe solid solutions. Manganese, a group VII element, is here, just as cadmium, a divalent positive ion. The ionic radius of Mn^{2+}, similar to that of Cd^{2+}, is somewhat smaller than the Hg^{2+} radius, and manganese ions replace mercury ions in the HgTe lattice. $Hg_{1-x}Mn_xTe$ crystals with x up to $x \simeq 0.35$ possess the zincblende structure (alloys with $x > 0.35$, just as MnTe, have a hexagonal lattice). Te atoms form one face-centered cubic sublattice, whereas Hg and Mn atoms are more or less randomly distributed in the other sublattice. The s-symmetry band is formed from $6s^2$-electrons of Hg and $4s^2$-electrons of Mn, the p-symmetry band from p-electron of Te.

The energy gap $\mathcal{E}_g$ between the s- and p-bands in $Hg_{1-x}Mn_xTe$ varies practically in a linear fashion as the Mn content is varied up to $x \leq 0.02$. For $x \simeq 0.07$ the GS turns into a semiconductor with a gap $\mathcal{E}_g > 0$. Investigations into the electric and optical properties of SMS HgMnTe in the absence of a magnetic field (or at sufficiently high temperatures) have shown that the aforementioned properties differ little or do not differ at all from those of HgCdTe-type crystals.

The situation changes dramatically when an SMS is placed in a magnetic field. SMS contain transition element atoms with unfilled electronic 3d-shells. The exchange interaction of band s- or p-electrons with 3d-electrons localized in Mn ions[27)] causes a considerable reconstruction of the energy spectrum of band electrons in a magnetic field. In turn, this gives rise to a number of peculiarities in the transport and magnetooptical phenomena at low temperatures. Thus the transverse magnetoresistance in GS HgMnTe first rises abruptly with increasing magnetic field and then falls to the same

27)The outer shell of the MN atoms is constituted by 4s-electrons, and the 3d-electrons form the inner shell.

extent, whereas the longitudinal magnetoresistance decreases monotonically (Davydov et al., 1980); the Shubnikov-de-Haas oscillation maxima in HgMnSe split up with increase in temperature, and the oscillation amplitudes vary nonmonotonically with temperature; the magnetoabsorption spectra turn out to be sensitive to temperature, etc. (Jaczynski et al, 1978; Bastard et al., 1977, 1978).

The fact that manganese 3d-electrons possess a noncompensated magnetic moment naturrally affects the magnetic properties of HgMnTe. Magnetic susceptibility, heat capacity and electron paramagnetic resonance studies have shown that for SMS HgMnTe and HgMnSe in a certain region of transition element Mn concentrations, when the temeprature is lowered below some T_s, the crystal passes from the paramagnetic to the spin-glass phase[28)] (Nagata et al., 1980).

The GS HgTe is a diamagnet, and MnTe an antiferromagnet. As the Mn content of $Hg_{1-x}Mn_xTe$ is increased the direct exchange interaction between the magnetic moments of Mn^{2+} ions enhances, as a result of which the magnetization increases. At Mn^{2+} ion concentrations $x \lessapprox 0.16$ the magnetization M in weak magnetic fields satisfying the condition $\bar{g}\mu_B SH \ll k_B T$ (for HgMnTe this corresponds to $H \lessapprox 60$ kOe) increases in proportion to H, as in typical paramagnets:

$$\vec{M} = \chi \vec{H}.$$

The paramagnetic susceptibility χ at a small Mn^{2+} ion content ($x \lessapprox 5 \times 10^{-3}$) is described by the Curie law

$$\chi = \frac{N_m S(S+1)(\bar{g}\mu_B)^2}{3k_B T} = \frac{\text{const}}{T} \tag{3.37}$$

with $N = xN_o$ being the manganese ion concentration, N_o the number of unit cells, $\bar{g}$ the spectroscopic splitting factor of manganese d-electrons, $S = 5/2$ the quantum number of the total spin magnetic moment of a manganese ion.

In sufficiently strong magnetic fields for which $\bar{g}\mu_B SH \gg k_B T$, the magnetization M ceases to depend on H, and the saturation M is

$$M = N_m \bar{g} \mu_B \langle \hat{S}_z \rangle , \tag{3.38}$$

where $\langle \hat{S}_z \rangle$ is the thermodynamic average of the projection on the z axis of the spin operator $\hat{S}$ of a Mn^{2+} ion. In a magnetic field parallel to the z axis, it is only the z component of the mean value $\langle \hat{S} \rangle$ that remains: $\langle \hat{S} \rangle = \langle \hat{S}_z \rangle$; the other two components $\langle \hat{S} \rangle$ are equal to zero. The magnetic field and temperature dependence of S_z ,

$$\langle \hat{S}_z \rangle = SB_s\left(\frac{\bar{g}\mu_B SH}{k_B T}\right) , \tag{3.39}$$

28) The spin glass is the state of a disordered magnetic system below the temperature T_s at which the directions of the magnetic moments are "frozen" chaotically (at $T<T_s$ the directions of the moments remain invariable with the lapse of time). In the paramagnetic phase, by contrast with the spin-glass phase, the directions of the magnetic moments fluctuate in time quasicontinuously and in a random fashion.

determines the H and T dependence of the magnetization M. In the Equation (3.39) $B_s(\chi)$ is a Brillouin function which describes the M(H, T) relation in the absence of an exchange interaction among Mn^{2+} ions over the entire range of values of the parameter $\bar{g}\mu_B SH/k_B T$.

With the Mn^{2+} ion concentration increased above $x \approx 10^{-2}$ and at temperatures T > 15 K, the magnetic susceptibility of $Hg_{1-x}Mn_xTe$ is described not by the Curie law (3.37) but by the Curie-Weiss law

$$\chi = \frac{N_m S(S+1)(\bar{g}\mu_B)^2}{3k_B(T+\Theta)} = \frac{\text{const}}{T+\Theta} , \qquad (3.40)$$

where $\Theta > 0$. The temperature dependence of the susceptibility (3.40) testifies to the presence of an exchange interaction among Mn^{2+} ions.

A sensible influence on the spectrum of band p-electrons in SMS and, consequently, on the transport and magnetooptical properties thereof is exerted by the exchange interaction between p-electrons and 3d-electrons of manganese ions that are randomly distributed in the cationic sublattice of the crystal. In order to take into account the effect of the exchange interaction on the spectrum, it is necessary to add to the one-electron Hamiltonian the exchange Hamiltonian (Bastard et al., 1978; Jaczynski et al., 1978).

$$\mathcal{H}_{exch} = \sum_i I(\vec{r} - \vec{R}_i)\, \hat{s}\, \hat{S}_i , \qquad (3.41)$$

with $\hat{s}$ being the spin operator of a band electron, $\hat{S}_i$ the total-spin operator of a Mn^{2+} ion, and I the exchange interaction integral. The summation in this equation is performed only over the lattice sites i whose magnetic moment is non-zero. Since the operator $\mathcal{H}_{exch}$ possesses no translational lattice symmetry, one has to make recourse to an approximate representation that possesses crystal periodicity and allows this operator to be added to the Hamiltonian of the one-electron Schrödinger equation.

Band structure calculations for solid solutions such as $Hg_{1-x}Cd_xTe$, $Hg_{1-x}Mn_xTe$, a.s.f. are normally carried out to the so-called virtual crystal approximation. Instead of the random potential of Hg^{2+} and Cd^{2+} ions, some averaged potential is introduced:

$$V(\vec{r}) = (1-x)\, V_{Hg} + x V_{Cd} \qquad (3.42)$$

where (1-x) is the probability of finding a Hg^{2+} ion on a given lattice site, x the probability of finding a Cd^{2+} ion on the same site. In contradistinction to the potentials V_{Hg} and V_{Cd}, the potential (3.42) is a periodic one. An appropriate procedure of averaging over the metallic sublattice sites is carried out for the exchange interaction Hamiltonian (3.41). As a result of this averaging, the sum (3.41) is extended over all the sites of the metallic sublattice. Similar to the expression (3.42), a weighting appears in the Hamiltonian (3.41), which allows for the probability of finding a magnetic moment on a given site:

$$\mathcal{H}_{exch} = x s_z \langle \hat{S}_z \rangle \sum_i I(\vec{r} - \vec{R}_i) \tag{3.43}$$

with s_z being the z component of the spin operator eigenvalue $\hat{s}$. On averaging, the probability of finding a magnetic moment is the same for all sites and the Hamiltonian (3.43) possesses lattice periodicity.

As calculations shows, the correction for the Landau level energy due to the exchange interaction, on the assumption that the conduction band is isotropic and parabolic (Bastard et al., 1978; Lyapilin et al., 1980), is determined by the expressions

$$\Delta\mathcal{E}_{N,\uparrow\downarrow} = \frac{\beta N_m \langle \hat{S}_z \rangle}{6} \begin{cases} \frac{8N - 1}{4N + 1} & \text{for } s_z = +\frac{1}{2}, \\ \frac{9 - 8N}{4N + 3} & \text{for } s_z = -\frac{1}{2}, \end{cases} \tag{3.44}$$

where N is the number of the Landau level, $\beta = \int p_x^2 I(\vec{r}) d\vec{r}$ a parameter specifying the exchange interaction of conduction band p-electrons with manganese d-electrons, p_x the symmetry wavefunction component.

The Equation (3.44) tells us that the exchange corrections $\Delta\mathcal{E}_{N,\uparrow\downarrow}$ for two spin orientations $s_z = +1/2$ and $s_z = -1/2$ are of the same sign for a Landau level with $N = 0$ and are of different sign for levels with $N \gg 1$. When $N \gg 1$ the corrections $\Delta\mathcal{E}_{N,\uparrow\downarrow}$ do not depend on Landau level number. It is clear that the exchange interaction should affect the band electron g factor characterizing the spin splitting of levels in a magnetic field $\Delta\mathcal{E}_N = \mathcal{E}_{N,\uparrow} - \mathcal{E}_{N,\downarrow} = |g|\mu_B H$ (see Eq. (2.33)). This effect reduces primarily to the quantity g becoming a function of both magnetic field and temperature.

For the p-electrons of an isotropic parabolic conduction band, allowance for the exchange interaction leads to the following expression for the effective g factor g*:

$$g^* = g + \frac{\beta N_m \langle \hat{S}_z \rangle}{\mu_B H} \tag{3.45}$$

With neglect of the exchange, the exchange interaction parameter β and the g factor are of the same sign for HgMnTe crystals: $\beta < 0$ and $g < 0$. Therefore the spin splitting of Landau levels $|g^*|\mu_B H$ under the influence of the exchange interaction may only increase (Fig. 34a)). As a result of the effect of the exchange interaction, the g factor of band electrons $|g^*|$ for HgMnTe amounts to hundreds of unities, whereas for HgCdTe crystals, which are similar in the major parameters, it does not exceed a few decades. When the value of the second term involved in the Equation (3.45) is sufficiently large (in magnitude) the usual arrangement of the levels may be perturbed: The upper spin sublevel of the Landau level, say N=0, may turn out to be above the lower spin sublevel of the Landau level N=1 (Fig. 34b)), i.e., an inversion of spin sublevels belonging to different Landau levels may take place. The g*(T) dependence here will be monotonic since $\langle \hat{S}_z \rangle$ always decreases with increasing temperature. The above features peculiar to the behavior of spin sublevels should have effect on spin-dependent transport phenomena.

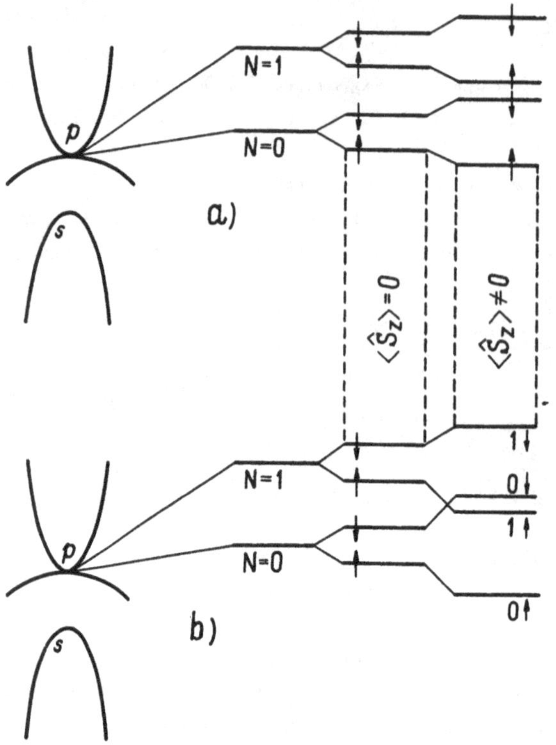

Fig.34. Schematic representation of the spin splitting of the first two Landau levels of the conduction p-band without allowance for the exchange effects ($<\hat{S}_z> = 0$) and with allowance for these ($<\hat{S}_z> \neq 0$): a) exchange interaction of moderate magnitude, b) strong exchange interaction

In studying Shubnikov-de-Haas oscillations of the magnetoresistance on SMS, peculiarities have been revealed which do not show up in semiconductors that are free from magnetic ions. We wish to enlarge upon some of these peculiarities (Jaczynski et al., 1978; Lyapilin et al., 1983). It has turned out that the oscillation amplitudes vary in a nonmonotonic fashion with temperature. Such a temperature dependence of amplitudes is typical of SMS only. In normal semiconductors this temperature dependence is monotonic and is used to determine charge carrier effective masses. With the magnetic field being fixed, the mass can be calculated from the ratio of amplitude A_1 at temeprature T_1 to amplitude A_2 at $T_2 > T_1$, which is always greater than unity. In SMS the amplitude ratio A_1/A_2 may, for different peaks, be both greater and less than unity. If we try to find the effective mass m for the ratio $A_1/A_2 < 1$, we will obtain m < 0. This means that the formula with the help of which $\tilde{m}$ is calculated from the quantity A_1/A_2 is inapplicable.

Very unusual from the point of view of the concepts about nonmagnetic semiconductors is the considerable displacement of oscillation peaks with temperature. In the case of strong electron gas degeneracy the peak position in normal semiconductors is independent of temperature. If the degeneracy is not sufficiently strong, i.e., if $\mathcal{E}_F$ exceeds k_BT by a small amount, the peaks shift somewhat toward lower fields. Apart from this, some peaks in SMS split up with increase in temperature, which has been found to come from the occurrence of a Landau level spin splitting. Nothing of the kind has been observed in nonmagnetic semiconductors. As the temperature is elevated split peaks in normal semiconductors become diffuse, whereas in SMS the splitting arises and enhances. All these features of the Shubnikov-de-Haas oscillations are due to the interaction of band electrons with transition element ions: The exchange additions to the band electron energy, as we have seen, depend on temperature and are responsible for all the temperature "anomalies" of the transport and magnetooptical phenomena.

To the two-band approximation and with neglect of the non-sphericity of the isoenergetic surfaces of p-electrons, the g factor for the lower Landau level (N=0) is (cf. Eq. (2.38))

$$g = 2(k - 3\bar{\gamma} - \frac{1}{3}x\beta < \hat{S}_z >) . \tag{3.46}$$

The exchange interaction between band and localized electrons brings about a very substantial change in the position of the valence-band heavy-hole levels in a magnetic field. The exchange interaction has a particularly strong effect on the position of the upper valence band Landau level relative to the conduction band edge, i.e., the lowest-lying electronic Landau level. A calculation shows that when the magnetic field is increased these levels may intersect, i.e., the valence band may overlap with the conduction band (Bastard et al., 1978). A zero-gap semiconductor thus turns into a semimetal.

The overlap energy of these levels will increase until the exchange addition to the valence level energy, proportional to the magnetization M, i.e., $< \hat{S}_z >$, attains saturation in a magnetic field. The further upward shift of the valence level on the energy scale with increasing magnetic field is now determined not by the magnetization, which has reached saturation, but by the field H dependence of the orbital energy of a heavy hole, $\hbar\omega_h = \hbar eH/m_h$. This leads to a decrease in overlap (Fig. 35).

It might be well to point out that the relative value of the exchange correction to the valence level energy $\Delta\mathcal{E}_{N,\uparrow\downarrow}/\hbar\omega_h$ is much weightier than the correction to the conduction band level $\Delta\mathcal{E}_{N,\uparrow\downarrow}/\hbar\omega_e$ since the orbital electron energy $\hbar\omega_e = \hbar eH/m_e$ exceeds the orbital hole energy $\hbar\omega_h$ by a small amount, because of the small effective mass m_e. In some magnetic field H_{cr} the overlap of the levels becomes equal to zero, and at $H > H_{cr}$ a semimetal turns into a semiconductor. It has been found experimentally that for $Hg_{1-x}Mn_xTe$ with x = 0.004 the semimetal-semiconductor transition at T = 2 K occurs in a field H_{cr} = 37 kOe (Bastard et al., 1978), and for $0.02 \le x \le 0.055$ in fields $80 \le H_{cr} \le 110$ kOe (Davydov et al., 1981).

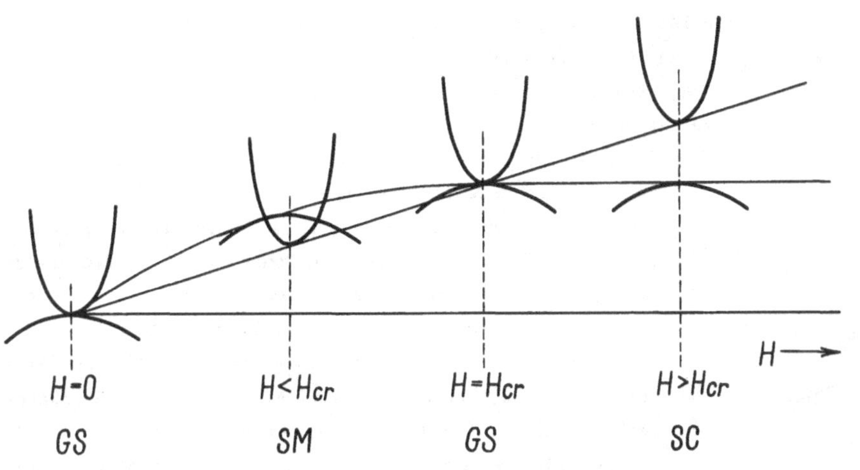

Fig. 35. Displacement of the upper valence level relative to the lower conduction band level in a magnetic field: GS - gapless semiconductor, SM - semimetal, SC - semiconductor

The same overlap effect is attained by increasing the manganese content of HgMnTe at constant temperature and in a constant magnetic field. With increasing Mn^{2+} ion concentration x the energy of the upper valence level rises, and at some value of x_{cr} this level intersects the lower conduction band level and a zero-gap semiconductor turns into a semimetal. The magnitude of band overlap depends not only on Mn^{2+} concentration but also on magnetic field and temperature. So peculiar an evolution of the energy levels with varying transition element content and magnetic field intensity naturally gives rise to a number of peculiarities in the kinetic coefficients of SMS.

A very unusual behavior of the magnetoresistance has been observed for GS $Hg_{1-x}Mn_xTe$ (Davydov et al., 1981). At helium temperatures the transverse magnetoresistance curves $\rho_{xx}(H)$ have been found to exhibit a maximum in magnetic fields $H = H_{max}$. The longitudinal magnetoresistance $\rho_{zz}(H)$ decreases with increase in H, starting from some rather weak field (Fig. 36). No such decrease of ρ_{zz} and ρ_{xx} (at $H>H_{max}$) with increasing H has been observed for HgCdTe with parameters close to those of the HgMnTe crystals investigated (which, however, contain no ions with a noncompensated magnetic moment!)

The decrease of ρ_{xx} and ρ_{zz} in a magnetic field may be explained rather naturally by the conduction and valence bands overlap discussed in the foregoing. When the valence and conduction band levels intersect as the field is increased, the concentration of electrons in the conduction band grows and, accordingly, the resistivity should decrease. But why does ρ_{xx} in fields $H<H_{max}$ increase with field, while there is no such growth for ρ_{zz}? The point is that with the current and the magnetic field being mutually perpendicular, the electron is deflected by the field. This deflection may be described as an effective decrease of the free path length and, therefore, of the electron mobility too. Consequently, the resistivity should increase. Fields $H \gtrsim 1$ kOe are strong for electrons ($\mu_e H>1$) and for 1 kOe$<H \lesssim H_{max} \approx$ 5-7 kOe the effect of decreasing mobility due to Larmor twisting is stronger than the effect of increasing electron concentration. At $H>H_{max}$ the contribution to the conductivity by the effect of increasing electron concentration becomes dominant, and the resistivity ρ_{xx} falls. In the longitudinal configuration, when the current is parallel to the magnetic field, the electrons do not undergo Larmor twisting and the resistivity ρ_{zz} starts decreasing with field as soon as a band overlap arises.

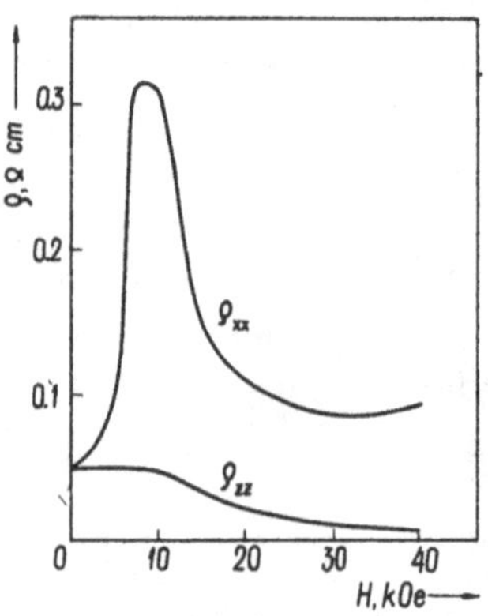

Fig.36. Variations of ρ_{xx} and ρ_{zz} in a magnetic field for a $Hg_{0.6}Mn_{0.4}Te$ sample at 4.2 K.

Gluzman, Sabirzyanova, and Tsidilkovski have found that the Hall coefficient R of the semimagnetic hole semiconductor $Hg_{0.88}Mn_{0.12}Te$ varies anomalously with temperature as compared to normal semiconductors. The specific resistivity of the sample investigated varied weakly in the 1.3-240 K range and was approximately 0.025 cm.

With the temperature decreased below 4.2 K the Hall coefficient R, which is positive at $T \approx 4.2$ K in rather weak magnetic fields $H<10^3$ Oe, decreases and, what is quite unexpected, becomes negative at $T < 2$ K. At high temperatures $T > 4.2$ K and in magnetic fields $H > 10^3$ Oe the R(H,T) dependences are typical for narrow-gap semimagnetic semiconductors with two species of charge carriers.

In weak magnemed fields a dominant contribution to transport phenomena is made by light conduction band electrons, and the Hall coefficient R is negative. As the magnetic field is increased (the measurements were performed in fields H up to 60 kOe and over the temperature interval 1.24-240 K) $|R|$ decreases. Let it be noted that for $T \lesssim 100$ K and $H \gtrsim 1 \times 10^4$ Oe the Hall coefficient R reverses sign since heavy valence band holes dominate under these conditions.

Altogether incomprehensible from the viewpoint of the theory of kinetic effects for "normal" p-type semiconductors is the reversal of sign in R from positive to negative when the temperature is decreased below 2 K and in magnetic fields $H < 10^3$ Oe. This unusual behavior of the Hall effect can be explained if one assumes that the investigated $Hg_{0.88}Mn_{0.12}Te$ crystals exhibit the spontaneous (anomalous) Hall effect which is typical of ferromagnetic metals and semiconductors. The magnetic induction B dependence of the Hall resistivity ρ_{xy} (off-diagonal magnetoresistance tensor component) is determined by the equation

$$\rho_{xy} = R_o B + R_s M , \qquad (3.47)$$

with R_o being the normal Hall effect coefficient, R_s the spontaneous Hall effect coefficient, M the magnetization of the sample. As can be seen from the Equation (3.47), a major feature of the spontaneous Hall effect is the dependence of the Hall resistivity ρ_{xy} (and, accordingly, of the Hall field E_y) on the magnetization of the sample.

Even in early investigations of the Hall effect in the ferromagnetic metals Fe, Co, and Ni it was found that the Hall field E_y is proportional to the angle of Faraday rotation of the electromagnetic wave polarization plane which, in turn, is proportional to the magnetization of the sample. Kikoin (1940) has shown that even in the paramagnetic region, i.e., at temperatures above the Curie temperature Θ, the Hall field in ferromagnets is determined mainly by the expression (3.47).

In the region of temperatures where $Hg_{0.88}Mn_{0.12}Te$ crystals exhibit the aforementioned Hall effect features, i.e., at $T \approx 2$ K, a phase transition from the paramagnetic to the spin-glass state occurs. Thus, when the $Hg_{0.88}Mn_{0.12}Te$ crystal passes to the spin-glass phase, a negative Hall field component appears which, when the temperature is lowered, is responsible for the decrease of the off-diagonal magnetoresistance tensor component ρ_{xy}. The idea suggest itself that this field component E_y is due to the spontaneous Hall effect, i.e., the term $R_s M$ involved in the expression (3.47). If this assumption is true, the measured Hall effect (resistivity ρ_{xy}) should depend on sample magnetization M.

To verify whether a correlation exists between the Hall resistivity ρ_{xy} and the magnetization M, the aforementioned authors staged the following two experiments. The dependence of the Hall resistivity ρ_{xy} on the magnetic induction B at temperature 1.33 K was measured. In one of the experiments the cooling of the sample to 1.33 K was done in the absence of a magnetic field, in the other the sample was cooled in a magnetic field with induction B = 10 kGs. The magnetization of the sample in these two experiments was different: In the case of cooling in a magnetic field the magnetization is larger than of the sample cooled in a zero field. Hall effect measurements have shown that the $\rho_{xy}(B)$ curves, too, differ substantially: Over the field interval $0 < B \lesssim 0.7$ kGs the absolute value of the Hall resistivity $|\rho_{xy}|$ in the sample cooled in a magnetic field with induction $B = 10$ kGs is larger than that measured when cooling the sample in the absence of a magnetic field. In the limit $B \to 0$ the Hall resitivity $|\rho_{xy}|$ of the sample cooled in the field $B = 10$ kGs is about twice as large as that for the sample cooled in field $B = 0$.

Thus the off-diagonal magnetoresistance tensor component has been found to correlate with the magnitude of sample magnetization M. This permits the conclusion that the unusual dependence of the Hall resistivity ρ_{xy} of the SMS $Hg_{0.88}Mn_{0.12}Te$ on magnetic field B and temperature T at $T < 2$ K is due to the spontaneous Hall effect arising in the spin-glass phase.

CONCLUSION

Practical Applications of Gapless Semiconductors

GS and narrow-energy-band semiconductors with $\mathcal{E}_g < 0.1$ eV possess a number of unique properties that open up vast prospects of their practical utilization. The entire spectrum of possible applications of these materials is based on the extremely high sensitivity of their band structure to external effects such as magnetic field, electromagnetic radiation, temperature, impurities, etc.

A most important area of application is the utilization of GS to create electromagnetic radiation detectors and generators encompassing a wide wavelength range - from radio to infrared waves. The principle of operation of such devices, manufactured until quite recently from normal wide-energy-band semiconductors, is based on the internal photoeffect. With the semiconductor being exposed to light of some frequency, the impurity-level or valence-band electrons are transferred into the conduction band. As a result, an extrinsic or intrinsic conductivity arises, and the net conductivity increases. Devices employing the phenomenon of increasing conductivity under the effect of radiation are called photoresistors. Measuring the conductivity permits detection of the radiation that has caused an internal photoeffect.

Photodetectors built on conventional semiconductors (Ge, Si, InSb, etc.) employ the extrinsic conductivity, whereas devices made involving the use of GS (HgCdTe, HgCdSe, HgZnSe, etc.) employ the intrinsic photoconductivity. The latter has a substantial advantage over the impurity photoconductivity. Since the intrinsic light absorption is considerably more effective than the extrinsic, a prescribed level of detectable or generable radiation power in the case of intrinsic photoconductivity is attained in a much smaller (by 3 or 4 orders of magnitude) bulk than in the case of extrinsic photoconductivity. This makes it possible to create many-element detectors as well as to raise the spatial resolution and directivity of radiation detection. Another advantage to photodetectors based on the use of intricsic conductivity is the possibility of employing these at higher temperatures, up to room temperature, by contrast with extrinsic -conductivity photodetectors whose operating temperatures lie below 20 K.

An indication that a radiation occurs in photoresistors is the alteration of the crystal's conductivity. The internal effect is also used in photodiodes, another variety of radiation detectors. But the radiation in photodiodes is registered by the photo e.m.f. that arises in the crystal. A photodiode is a semiconductor containing a p-n junction, i.e., a potential barrier between the hole (p) conductivity and electron (n) conductivity regions. Light-generated electrons and holes are separated spatially by a p-n junction giving rise to an e.m.f.; it is this e.m.f. which permits the radiation to be detected. The internal photoeffect occurring in alloys that contain GS is laid at the foundation of solar to electric energy converters and infrared-devices.

If what serves as a photodetector is a semiconductor with a fixed value of gap between the bands (intrinsic conductivity) or with a fixed gap between the impurity level and the band (extrinsic photoconductivity), it is possible to record the radiation of a determinate wavelength region. The upper bound of photoconductivity-inducing radiation wavelengths λ is determined by the ratio $2\pi\hbar/\lambda \geq \mathcal{E}_g$ which is equivalent to the inequality $\hbar\omega \geq \mathcal{E}_g$ (the energy of a light quantum of frequency ω should be not smaller than the energy gap, to bring about charge carrier generation). A photodetector based on GS has a tremendous advantage over a radiation detector made on the basis of a normal semiconductor: The gap value $\mathcal{E}_g$, i.e., the boundary radiation wavelength can be controled by means of external effects. This means that by varying the gap $\mathcal{E}_g$, radiation detectors and generators may be created on the basis of GS.

As is known, the gap value may be varied by applying pressure or a magnetic field. In practice, however, the energy gap $\mathcal{E}_g$ in materials used for instrument making is adjusted by alloying a GS with $\mathcal{E}_g < 0$ (for example, HgTe) and a semiconductor possessing a wide gap $\mathcal{E}_g > 0$ (for example, CdTe). Increasing the Cd content of HgCdTe alloy allows the gap between the conduction and valence bands to be varied between 0 (approximately 16% Cd) and 1.6 eV (CdTe). Detectors based on HgCdTe are sensitive over the entire infrared spectral range, but they are employed primarily for "atmospheric window" ranges, i.e., wavelength intervals where the atmosphere is transparent for infrared rays: from 8 to 14 and from 3 to 5 μm.

HgCdTe-based photodetectors are used both as photoresistors and as photodiodes. Photoresistors are used in the temperature range between 77 and 300 K, i.e., with and without cooling. Their major characteristics, viz., minimum detectable radiation power and the time during which the signal of the detector reaches a maximal value, as well as other characteristics are very high. In service properties, infrared radiation detectors made on the basis of HgCdTe alloys excel all the photodetectors known currently and are extensively used in infrared imaging (in medicine, environment control, prospecting for mineral resources).

Photodiodes have been used as direct solar to electric energy converters. Compounds of elements belonging to the second and sixth groups of the Periodic Table have proved most suitable for creating solar photocells, for the energy gap in them may be so chosen that a large portion of the solar emission spectrum is absorbed. Highly effective is the photocell built on the basis of the graded-gap structure[29)] of p-HgCdTe-n-HgCdTe in which the gap in the upper p-type layer decreases in the direction of the p-n junction and remains constant in the n-type layer.

The internal electric field in the p-type layer accelerates nonequlibrium (light-induced) charge carriers toward the p-n junction. This considerably lowers the recombination of charge carriers near the surface upon which light falls, and thus increases the

29) A graded-gap structure is one in which the energy gap varies in some direction. For example, a structure that forms by alloying $Hg_{1-x}Cd_xTe$ crystals with different but gradually varying Cd content.

photocell efficiency (the ratio of electric current energy to incident light energy). According to theoretical estimates, the efficiency of this structure incorporating a p-type and an n-type HgCdTe crystal should amount to 19%, whereas for a CdTe-based photodiode under similar conditions it is as low as 11%.

The internal photoeffect has recently found important applications in a new type of devices - charge-coupled devices. These devices, constituting a three-layer metal-indulator-semiconductor structure, operate on a rather simple principle. Under the radiation penetrating into the semiconductor through thin metal and insulator layers, an electron-hole pair is produced. The electric field applied to the structure, which is a capacitor, extends (destroys) the electron-hole pair : The positive charge is localized (bound) near the negative metallic electrode (but is separated from the latter by a thin layer of insulator), the negative charge is "withdrawn" by the field from the semiconductor. The magnitude of the bound charge is proportional to the absorbed radiation intensity, and this charge may persist in the three-layer cell as long as desired. With the external electric field removed from the capacitor, the charge that has piled up is injected into the bulk of the semiconductor and an electric current arises in the external circuit, which may be recorded.

When fitted in staggered rows on the same semiconductor plate, such photodetectors form a matrix that is capable of converting light to electric signals which, in turn, may be transformed into visible images by means of conventional television receiving sets. The techniques and procedures of fabricating suchlike matrices have been brought to so high a level that image-sensing electronic devices disposed on a single semiconductor plate can be created.

Currently, a major trend in the development of infrared-range image detectors is to devise matrix receivers that employ the fundamental (intrinsic) radiation absorption (which, as already stated, has great advantages over the extrinsic absorption) in GS-based narrow-gap semiconductors.

For measuring magnetic fields and in the various electronic devices, Hall generators and magnetoresistors are extensively used. A Hall generator is a semiconductor plate which is placed in a magnetic field perpendicular to the current flowing in the plate. From the Hall e.m.f. that arises, one can determine with high accuracy the magnetic field intensity. In magnetoresistors the principle of the change in resistance in a magnetic field is employed. For semiconductors with high charge carrier mobilities this change is large, which allows magnetoresistors to be used for precision measurements of the magnetic field.

Advantages to Hall generators and magnetoresistors are their speed of response,their small dimensions, and the absence of moving parts. In their quality, Hall generators made from HgCdTe are not inferior to the best InSb-based semiconductor sensors. And magnetoresistors made on the basis of GS $Hg_{0.9}Cd_{0.1}Te$, in which the electron mobility is high, not only may compete with but also excel, in some parameters, the best devices based on InSb with an admixture of Ni.

GS $Hg_{1-x}Cd_xTe$ are used as materials for thermojunctions and coolers. The thermoelectric and thermomagnetic effectiveness of these materials is very high: It exceeds by one or three orders of magnitude the effectiveness of semiconductor materials thus far employed to this end. At low temperatures (50-90 K) thermojunctions and coolers based on GS possess high efficiencies.

Some Results and Prospects

Over the three decades that have passed since the discovery in HgTe of an anomalously small gap between conduction and valence bands (1955-1957) and since the validation of the inverse band structure of α-Sn (1963) many laboratories of the world have been intensively studying the properties of GS. Efforts by a large number of researchers have enabled many "secrets" to be disclosed and the "anomalous" behavior of these materials to be understood. A number of experiments have reliably confirmed the inverse band order in HgTe, α-Sn, HgSe, HgCdTe, HgCdSe, HgMnTe, etc. The influence of many-electron effects on the band structure has been theoretically analyzed. Some experimental results are available which bear testimony to the fact that the electron-electron interaction exerts influence on the energy spectrum. The major band parameters of the GS studied have been determined.

A great deal of attention has been given to the electron spectrum rearrangement and, specifically, the GS-semiconductor electronic transition under the effect of pressure, magnetic field, and temperature. A very intensive study has been made of the influence of substitutional impurities and other imperfections on the energy spectrum and transport and magnetooptical effects. Of great interest are GS that contain transition element impurities with noncompensated magnetic moments. This interest comes from the fact that the exchange interaction of conduction electrons with these impurities alters the band picture substantially.

We have discussed experimental and theoretical research results which bear primarily on HgTe, HgCdTe, and, to a somewhat lesser extent, on α-Sn, HgMnTe, and other GS. This is so because HgTe and solid solutions HgCdTe have currently received the most study. However, the conclusions drawn for these materials on the strength of experimental and theoretical work may undoubtedly be used to interpret physical properties of other GS as well.

Of fundamental importance is the problem of investigating the effect of the electron-electron interaction on the spectrum in the vicinity of the degeneration point of the conduction and valence bands in GS. The vicinity of this point has thus far been inaccessible since in the purest HgTe crystals the electron concentration (determined by the donor concentration) and, accordingly, the Fermi energy are too high. The following important circumstance, which is a serious obstacle to the study of the immediate vicinity of degenerate bands, must be noted here.

Any perturbation of the lattice of a perfect crystal, whether departures from stoichiometry or nonuniformities in the composition of solid solutions, whether impurities or even electrically inactive defects which are of a random character, leads to overlap of the valence band with the conduction band. As a result, the charge carrier density of states of GS turns out to be similar to the counterpart in a semimetal, i.e., is nonzero whatever the temperature may be. Since the hole effective mass m_h exceeds the electron effective mass m_e by a large amount, the Fermi level at low temperatures lies in the conduction band. In consequence, the band electron concentration remains finite to T = 0 K (except for the case of a heavily doped p-type crystal, $N_A \gg N_D$).

No unambiguous answer can at present be given to the question whether the smallest Fermi energy values $\mathcal{E}_F \geq 1$ meV in GS are due to the complete ionization of residual donors with concentrations $N_D \approx 10^{15}$ cm^{-3} or arise from the effective overlap of the conduction and valence bands. Even if it becomes possible in future to make some progress in the technology of purifying single crystals and to lower the donor concentration by one or two orders of magnitude (although more than twenty-five years of experience in technological research of HgTe in laboratories of different countries show that no realistic ways are seen of bringing N_D and N_A below $\approx 10^{15} cm^{-3}$), the region of energies near the degeneracy point of the bands $\mathcal{E} \ll 1$ meV will remain inaccessible as before: The electrons possessing this energy are too deep under the Fermi level and do not participate in transport processes. In this context is unlikely that many of the predicted theoretical effects arising from electron spectrum renormalization due to electron-electron interaction may, in the near future, be disclosed for HgTe or HgCdTe crystals.

A particularly strong density of states perturbation arises from the electric field of charged impurities in heavily doped and highly compensated crystals. The effective overlap of the conduction and valence bands under these conditions reaches ≈ 10 meV. Although no rigorous theory exists so far that permits description of the density of states for GS with high impurity concentration, it may be hoped that the estimates made on the basis of the theory of normal semiconductors will hold good.

Experimental research results show that the conduction band in GS such as HgTe, in contrast to the valence band, is practically insensitive to perturbations caused by crystal lattice imperfections. Thus the electron effective mass and the density of states of conduction band electrons do not differ from the counterparts for an ideal crystal. This allows the Fermi level position with respect to the degeneracy point of the bands in doped and compensated p-type GS to be determined very exactly, which is rather difficult to do for semiconductors with a gap. Thus GS offer a unique opportunity of tracing the variation of the Fermi level position as a function of impurity concentration and the degree of compensation in the vicinity of the Mott-Anderson transition. The conduction band here acts as a peculiar indicator that permits determination of the boundary between localized and delocalized states.

In the physics of the gapless state there are a number of interesting and fundamentally important problems that remain to be ascertained. One of the most urgent problems is to investigate features peculiar to and the stability of the electron spectrum near the degeneracy point of the bands and to ascertain the role of many-electron effects. We have already said that solution of this problem is deferred to a time when sufficiently perfect, nearly ideal crystals will be created. In this context it is now particularly important to concentrate effort on perfecting the technology of making high-purity single crystals and films of GS.

The problem remains unvolved concerning the contribution of impurity band charge carriers to transport processes. Specifically, the cause for the occurrence of a particular sign of the Hall effect due to acceptor band charge carriers is not clear. Nor is it clear why the mobilities of these carriers reach such large magnitudes.

In GS with acceptor concentrations $N_A < 10^{17} cm^{-3}$, localized impurity band states are hybridized with delocalized conduction band states. The undoubtedly interesting problem as to the influence of this hybridization on the character of the Mott-Andersond transition has not yet been investigated at all.

Some problems concerning the properties of GS with low acceptor content $N_A \ll N_M \lessapprox 10^{17} cm^{-3}$ remain to be ascertained. On the one hand, acceptors under these conditions remain neutral at sufficiently low temperatures, as indicated by the finite conductivity activation energy. This confirms the picture of the density of states with a sharp local peak on the background of the continuous conduction band spectrum (see Fig. 27a), b)). On the other hand, a resonance scattering of band electrons of energy ε_A on acceptors should occur in this case. This resonance scattering may be expected to lead to a considerable decrease in electron mobility at energies ε equal to acceptor level energy ε_A. However, no conclusive evidence has thus far been available for such an effect.

It appears tempting to comprehensively and profoundly investigate electrophysical properties of the only atomic (one-component) zero-energy-gap semiconductor known - gray tin. It remains an unfortunate fact that the body of experimental evidence concerning transport and other effects in gray tin is extremely scanty - there are literally only a few papers available. At the same time it is quite clear that the problem concerning the influence of defects, particularly such as departures from stoichiometry and vacancies, on the physical properties is ruled out entirely (or almost entirely, when vacancies are concerned).

The needs of radio-electronics have stimulated advances in commercial applications of narrow-gap semiconducting crystals based on mercury telluride. However, the GS is so interesting an entity from the standpoint of fundamental solid state physics that further attempt at studying any representative of this comparatively new family of materials seem justified.

We conclude by discussing some prospects of practical utilization of materials that contain GS as the basis. First of all, it should be noted that the extremely high electron mobility in GS, which exceeds the counterpart in classical semiconductors by 2-4 orders of magnitude, allows the speed of response of electronic devices to be increased considerably and to be brought to a level of hundreds of picoseconds.

Of indubitable interest is the use of GS in creating an infrared laser readjusted by a magnetic field. Very promising is the creation of very sensitive infrared radiation detectors in the range up to 1000 μm (1mm), i.e. , low-temperature (down to ≈ 10 K) thermal-radiation detectors.

Highly promising are infrared devices based on zero-gap and narrow-gap solid solutions of lead and tin chalcogenides with impurities of group III elements (In and Ga). These devices should possess fixed storage and large spatial resolution in optical recording of information. The In^+ ion and the neutral In atom are in potential wells of different depth. Therefore when the electron is coupled to the In^+ ion or the In atom is ionized, the electron should overcome some potential barrier U. At low temperatures $k_BT \ll U$ (experimentally, this is the range ≈ 4-50 K) the probability of overcoming the barrier is proportional to $\exp(-U/k_BT)$. This is what is responsible for the exponential dependence of the relaxation time τ of the number of band electrons on temperature: $\tau = \tau_0 \exp(T_o/T)$, where $\tau \approx 5 \times 10^6$s, $T_o \approx$ 200-250 K.[30] At $T \lesssim 4$ K the time τ amounts to several hours, i.e., the In atom is a fixed metastable trap for electrons whose concentration decreases slowly with the time. The energy level that corresponds to the In impurity is resonant, i.e., lies in the conduction band. At low temperatures the Fermi level lies above the In impurity level, and the electrons migrate very slowly from conduction band states with energies $\varepsilon > \varepsilon_{In}$ to the indium impurity level of energy ε_{In}.

30) At $T < 4$ K the major mechanism of the passage of electrons through a barrier is by tunneling, and the temperature dependence $\tau(T)$ will be somewhat different.

The relaxation time τ is so large that the electrons are sort of frozen in the conduction band. At high temperatures $k_B T \gtrapprox U$ the process of coupling of the electron to the In^+ ion or of In atom ionization occurs rapidly and the relaxation time τ is sharply accelerated. Thus for $T \simeq 2$ K the relaxation time τ is larger than 2 hours, and for $T \simeq 25$ K it is less than 0.1 s.

As already noted in the foregoing, $Hg_{1-x}Cd_xTe$ crystals with $x = 0.2$ possess an energy gap $\mathcal{E}_g \approx 0.1$ eV which corresponds to the wavelength of maximum thermal radiation at room temperature ($\lambda = 10$ µm). Therefore such crystals may be successfully employed to measure small temperature gradients in environment and in medicine. The high thermoelectric effectiveness of a device based on HgCdTe may be used to cool optical information converters, radiation detectors, amplifying output stages that heat in electronic devices, etc.

Solid solutions $Hg_{1-x}Cd_xTe$ show promise of success from the viewpoint of multicolor infrared photodetectors, as the spectral characteristic can be varied over a wide range by altering the composition of the solid solution. Multicolor photodetectors may be constructed in the following way. HgCdTe crystals with different Cd content (i.e., with different $\mathcal{E}_g$) and, accordingly, different regions of spectral sensitivity may be arranged in succession. Each short-wave detector (i.e., a detector with a larger gap $\mathcal{E}_g$) serves simultaneously, for longer-wave detectors, as a filter that truncates the short-wave spectral region. Thus all detectors receive radiation of different frequencies (different colors) from the same spatial region. By applying signals from detectors to relevant comparison circuits, a "thermal" chart of a radiating object may be drawn up.

GS may find applications as active elements of submillimeter and far-infrared waveguides controled by magnetic and electric fields or temperature. Here belong radiation amplitude modulators, phase inverters, switches, attenuators (devices for reducing the radiative power). GS open up an interesting possibility of creating radiation frequency analyzers on the basis of Fabry-Perot resonators tuned by magnetic field (instruments for Fourier analysis in the submillimeter spectral region). The resonator optical length may be readjusted with the help of the magnetic field, which alters the refractive index of radiation, rather than mechanically, by displacing the corresponding parts, as is normally done.

Thus GS currently enjoy wide use and are highly promising in creating controled electronic devices which will substantially extend the capabilities of radio- and opto-electronics in the yet unconquered spectral region of electromagnetic radiation at the boundary of optical and radio frequency ranges.

REFERENCES

Ablyazov, N.A.; Reich, M.E.; Efros, A.L., 1983, Zh.eksp.teor. Fiz. Pisma 38,103.

Abrikosov, A.A. 1974, Zh.eksp.teor.Fiz., 66, 1443; 1975, J. Low Temp.Phys. 18, 185.

Abrikosov, A.A.; Beneslavsky, S.D. 1970, Zh.eksp.teor.Fiz. 59, 1280.

Anderson, P.W., 1958, Phys.Rev. 109, 1492.

Arapov, Y.G.; Davydov, A.B.; Tsidilkovski, I.M.; Sheblushinina, N.G., 1982, Phys.Stat. Sol.(b) 110, 619.

Arapov, Y.G. Davydov, A.B.; Zvereva, M.L.; Stafeev, V.I.; Tsidilkovski, I.M., 1983, Fiz.Tekn.Poluprovodn. 17, 1392.

Arapov, Y.G.; Ponikarov, B.B.; Tsidilkovski, I.M.; Nesmelova, I.M., 1979, Fiz.Tekn.Poluprovodn. 13, 695.

Bastard, G., 1977, Phys.Stat.Sol.(b) 80,641.

Bastard, G.; Guldner, Y.; Mysielski, A.; Nozières, P.; Rigaux, C., 1974, Proceedings of the Twelfth International Conference on the Physics of Semiconductors, Stuttgart, edited by Pilkuhn. B.G. Teubner, Stuttgart, p.1162.

Bastard, G.; Rigaux, C.; Guldner, Y.; Mycielski, J.; Mycielski, A., 1978, J.d.Phys. 39, 87.

Bastard, G.; Rigaux, C.; Mycielski, A., 1977, Phys.Stat.Sol.(b 79, 585.

Berchenko, N.N.; Pashkovskii, M.V., 1976, Usp.fiz.Nauk 119, 223.

Bloch, F.; 1928, Zs.Physik 52, 555.

Brandt, N.B.; Belousova, O.N.; Ponomarev, Y.G.; Anderson,J.R., 1976, J.Low Temp.Phys. 24, 471.

Brandt, N.B.; Moschalkov, V.V., 1984, Adv.Physics 33, 193.

Brandt, N.B.; Tsidilkovski, I.M.; Moschalkov, V.V.; Ponikarov, B.B.; Scrbek, L.; Taldenkov, A.N.; Chudinov, S.M., 1983, Fiz. tverd.Tela 17, 18.

Brillouin, L., 1930, Compt.Rend. 191, 292.

Bush, G.; Kern, R., 1960, Solid State Physics, Vol.11 (New York), p.1.

Carvalho, M.; Fau, C.; Averous, M., 1982, Proceedings of the Sixteenth International Conference on the Physics of Semiconductors, Montpellier, edited by M. Averous. North Holland, Amsterdam. p.431.

Davydov, A.B.; Ponikarov, B.B.; Tsidilkovski, I.M., 1980, Phys.Stat.Sol.(b), 101, 127; 1981, Fiz.Tekn.Poluprovodn. 15, 881.

Dietl, T.; Dubowski, J.J.; Szymanska, W., 1978, Physics of Semiconductors, Inst.Phys.Conf. Ser. No.43. The Institute of Physics, Bristol. p.245.

Dornhaus. R.; Nimtz, G., 1976, Solid State Physics, Vol.78 Springer-Verlag: New York.

Dornhaus, R.; Nimtz, G.; Schlicht, B., 1983, Narrow-Gap Semiconductors. Springer Tracts in Modern Physics, No.98, Berlin, Heidelberg, New York.

Elliott, C.T.; Melngailis, J.; Harman, T.C.; Kafalas, J.A.; Kernan, W.C., 1972, Phys.Rev. B5, 2985.

Erdman O., 1951, J.Prakt.Chemie 52, 428.

Ewald, A.W.; Tufte, O.N., 1958, J.Appl.Phys. 29, 1007.

Finck, C.; Otmezguine, S.; Weill, G.; Vérié, C., 1972, Proceedings of the Eleventh International Conference on the Physics of Narrow Gap Semiconductors, Warsaw, Vol.2, PWN Polish ScientificPublishers, Warsaw, p.944.

Furdyna, J.K., 1982, J.Vac.Sci.Technol. 21, 220.

Galazka, R.R., 1970, Phys.Lett. A 32, 101.

Gelmomt, B.L., 1975, Fiz.Tekn.Poluprovodn. 9, 1912.

Gelmont, B.L.; Dyakonov, M.I., 1972, Zh.elsp.teor.Fiz. 62, 713.

Gelmont, B.L.; Ivanov-Omskii, V.I.; Kolomiets, B.T.; Ogorodnikov, V.K.; Smekalova, K.P., 1971, Fiz.Tekn.Poluprovodn. 5, 266.

Gelmont, B.L.; Ivanov-Omskii, V.I.; Tsidilkovski, I.M., 1976a, Usp.fiz.Nauk 120, 337.

Gelmont, B.L.; Ivanov-Omskii, V.I.; Konstantinova, N.N.; Mashovetz, D.V.; Parfeniev, R.V.; Yassievich, I.N., 1976b, Zh.eksp.teor. Fiz. 71, 1572.

Giriat, W.; Neifield, E.A.; Tsidilkovski, I.M., 1975, Fiz.Tekn. Poluprovodn. 9, 188.

Giriat, W.; Neifield, E.A.; Tsidilkovski, I.M.; Gorodilov, N.A., 1976, Fiz.Tekn.Poluprovodn. 10, 63.

Gluzman, N.G.; Sabirzyanova, L.D.; Tsidilkovski, I.M.; Gavaleshko, N.G.; Frasunyak, V.M., 1981, Fiz.Tekn.Poluprovodn. 15, 1647.

Gorodilov, N.A.; Neifield, E.A.; Harus, G.I.; Tsidilkovski, I.M., 1980, Fiz.Tekn.Poluprovodn. 14, 2357.

Groves, S.H., 1971, J.Phys.Chem.Solids 32, Suppl.1, 447.

Groves, S.H.; Brown, R.N.; Pidgeon, C.R., 1967, Phys.Rev. 161, 773.

Groves, S.H.; Paul, W., 1963, Phys.Rev.Lett. 11, 194; 1964, Proceedings of the Seventh International Conference on the Physics of Semiconductors, Paris. Dunod, Paris, p.41.

Groves, S.H.; Pidgeon, K.; Ewald, A.; Wasner, R., 1969,Proceedings of the Ninth International Conference on the Physics of Semiconductors, Leningrad, Vol.1, p.46, Nauka; Leningrad, 1970, J.Phys.Chem.sol. 31, 2031.

Guldner, J.; Rigaux, C.; Mycielski, A.; Couder, Y., 1977, Phys.Stat.Sol.(b) 81, 615.

Halperin, B.J.; Rice, T.M., 1968, Rev.mod.Phys. 40, 775.

Harman, T.C., 1967, Physics and Chemistry of II-VI Compounds. Wiley, New York, p.784.

Harman, T,C.; Logan, M.J.; Goering, H.L., 1958, J.Phys.Chem.Sol. 7, 228.

Harman, T.C.; Melngailis, J., 1974, Applied Solid State Science. Academic Press, New York, London, p.1.

Harus, G.I.; Tsidilkovski, I.M.; Shelushinina, N.G., 1983, Fiz. Tekn.Poluprovodn. 17, 1944.

Herman, F., 1955, Electronics 1, 103.

Herman, F.; Kortum, R.; Kuglin, C.; Van Dyke, J.; Skillman, S., 1968, in: Methods in Computational Physics, Vol.8, New York.

Ivanov-Omskii, V.I.; Mehtiev, A.Sh.; Rustambekova, S.A.; Ukraintsev, E.N., 1983, Phys. Stat.Sol. (b) 119, 159.

Jasynski, M.; Kossut, J.; Galazka, R.R., 1978, Phys.Stat.Sol.(b) 88, 73.

Joos, B.; Das, A.k.; Wallace, R.R., 1978, Phys.Rev. B18, 5691.

Kane, E.O., 1957, J.Phys.Chem.Sol. 1, 249; 1979, Proceedings of the International Summer School on the Physics of Narrow Gap Semiconductors, Nimes Springer-Verlag, Berlin, Heidelberg, New York, p.13.

Keldysh, L.V.; Kopaev, Y.V., 1964, Fiz.tverd.Tela 6, 2791.

Knox, R.S., 1963, Solid St.Phys. Suppl. 5, 100.

Liu, L.; Brust, D., 1968a, Phys.Rev.Lett. 20, 561; 1968b, Phys.Rev. 173, 777.

Long, D.; Schmit, J.L., 1970, Semiconductors and Semimetals, Vol.5, edited by R.K. Willardson and A.C. Beer, Academic Press, New York London, p. 175.

Luttinger, J.M., 1956, Phys.Rev. 102, 1030.

Lyapilin, I.I.; Karyagin, V.V., 1980, Fiz.tverd.Tela 22, 2859.

Lyapilin, I.I.; Tsidilkovski, I.M., 1985, Usp.fiz. Nauk 146, 35.

Lyapilin, I.I.; Ponomarev, A.I.; Harus, G.I.; Gavaleshko, N.P.; Maryanchuk, P.D., 1983, Zh.eksp. teor.Fiz. 8, 1638.

Mauger, A.; Friedel, J., 1975, Phys.Rev. B12, 2412.

Mycielski,J., 1982, Proceedings of theInternational Conference on High Magnetic Fields in Semiconductor Physics, Grenoble, p.431.

Naguta, S.; Galazka, R.R.; Knattak, C.D.; Mullin, D.P.; Akbarzaden, H.; Furdyna, J.K.; Keesom, P.H., 1980, Phys.Rev. B22, 3331.

Neuringer, L.J., 1968, Proceedings of the Ninth International Conference on the Physics of Semiconductors, Moscow, Vol.2 Nauka, Leningrad, p.757.

Overhof, H., 1971, Phys.Stat.Sol.(b) 43, 221.

Piotrzkowski, R.; Porowski, S.; Dziuba, Z.; Ginter, J.; Giriat, W.; Sosnowski, L., 1965, Phys.Stat.Sol. 8, K135.

Pollak, F.H.; Cardona, M.; Higginbotham, C.W.; Herman, F.; Van Dyke, J.P., 1970, Phys.Rev. B2, 352.

Porowski, S.; Zakrzewski, T., 1965, Phys.Stat.Sol. 11, K39.

Raymond, A.; Aulombard, R.L.; Konczewicz, L.; Bousquet, C.; Robert, J.L.; Royer, M., 1980, J.Phys.Soc.Japan 49, Suppl.A, 301.

Raymond, A.; Robert, J.L.; Aulombard, R.L.; Bousquet, C.; Valassiades, O.; Royer, M. 1981, Proceedings of the Fourth International Conference on the Physics of Narrow Gap Semiconductors, Linz, edited by E. Gornik, H. Heinrich and L. Palmetshofer, Springer-Verlag, Berlin, Heidelberg, New York, p.387.

Reynold, R.A.; Brau, M.J.; Kraus, H.; Bate, R.T., 1971, J.Phys.Chem.Solids 32, Suppl. 1, 511.

Robert, J.L.; Raymond, A.; Aulombard, R.L.; Bousquet, C., 1980, Phil.Mag. B42, 1003.

Rodot, H.; Triboulet, R., 1962, C.r.hebd.Séanc.Acad.Sci., Paris, 254, 852.

Roman, B.J.; Ewald, A.W., 1968, Bull.Am.Phys.Soc. 13, 408.

Roman, B.J.; Ewald, A.W., 1972, Phys.Rev. B5, 3914.

Schmit, J.L.; Stelzer, E.L., 1978, J.Electron.Mater. 7, 65.

Scott, W.; Hager, R.J., 1971, J.appl.Phys. 42, 803.

Scott, W.; Stelzer, E.L.; Hager, R.J., 1976, J.appl.Phys. 47, 1408.

Shklovskii, B.I.; Efros, A.L., 1979, Electron Properties of Doped Semiconductors, Nauka, Moscow.

Shubin, S.P.; Vonsovski, S.V., 1934, Proc.Roy.Soc. A 145, 159.

Sologub, V.V.; Ivanov-Omskii, V.I.; Muzhdaba, V.M.; Shalyt, S.S., 1971, Fiz.tverd.Tela 13, 1738.

Stankiewicz, J.; Giriat, W.; Bienenstock, A., 1971, Phys.Rev. B4, 4465.

Strauss, A.J.; Brebrick, R.F., 1970, J.Phys.Chem.Sol. 31, 2293.

Tsidilkovski, I.M., 1955, All-Union Conference on the Physics of Semiconductors, Leningrad (Leningrad: Nauka); 1957, Zh.tekn.Fiz., 27 1744; 1982, Band Structure of Semiconductors, Pergamon Press, Oxford.

Tsidilkovski, I.M.; Harus, G.I.; Shelushinina, N.G., 1985, Adv.in Physics 34, 43.

Tsidilkovski, I.M.; Schennikov, V.V.; Gluzman, N.G., 1982, Fiz.tverd.Tela 24, 2658; 1983, Fiz.Tekn.Poluprovodn. 17, 958.

Tuchendler, J.; Grynberg, M.; Couder, Y.; Thome, H.; Touless, R.L., 1973, Phys.Rev. B8, 3884.

Vydyanath, H.R., 1981, J.Electrochem.Soc. 128, 2609.

Vydyanath, H.R.; Donovan, J.C.; Nelson, D.A., 1981, Electrochem.Soc. 128, 2625.

Yafet, Y.; Keyes, R.W.; Adams, E.N., 1956, J.Phys.Chem.Solids 1, 137.

Semiconductor Surfaces and Interfaces

Their Atomic and Electronic Structures

by
Friedhelm Bechstedt / Rolf Enderlein

1988. 441 S., 149 Abb., 51 Tab.,
17 cm x 24 cm.
DDR 88,-- M; Ausland 88,-- DM
ISBN 3-05-500479-5
Bestell-Nr. 763 796 5
Bestellwort: Bechstedt/Enderlein 9091

"God created the solid state bulk, and the Devil made its surface". There are enough indications that Wolfgang Pauli, the great Swiss theoretician who made this remark, was indeed right. Today surface and interface physics is one of the most important and fundamental areas of solid state research. The achievements in this area are not only important for basic science, but also for microelectronics and optoelectronics which just have started to exploit the unique properties of abrupt semiconductor interfaces.
This book deals with the most fundamental properties of semiconductor surfaces and interfaces being their atomic and electronic structures. Properties which like the Schottky barrier heights of metal-semiconductor interface can be derived from these structures are also included. The book provides a thourough introduction to the subject which can also be understood by readers who have no special background in surface physics. It reviews the state of the art of the field on an advanced level covering also very recent results (published before 1987). The authors address all those who are interested in the modern physics of solid state surfaces and interfaces.

„Das Volumen der Festkörper schuf Gott, ihre Oberfläche wurde vom Teufel gemacht". Es gibt genügend Hinweise, daß Wolfgang Pauli, der große Schweizer Theoretiker, mit dieser Bemerkung recht hatte. Heute ist die Physik der Festkörperoberflächen und -grenzflächen eines der bedeutendsten und grundlegendsten Gebiete der Festkörperphysik. Die Ergebnisse auf diesem Gebiet sind nicht nur für die Grundlagenforschung, sondern auch für die Mikro- und Optoelektronik von Bedeutung, die gerade begonnen haben, die einzigartigen Eigenschaften abrupter Halbleitergrenzflächen zu nutzen.
Das Buch beschäftigt sich mit den grundlegendsten Eigenschaften der Halbleiteroberflächen und -grenzflächen, ihrer atomaren und elektronischen Struktur. Behandelt werden auch Eigenschaften, die — wie die Schottky-Barrierenhöhen — aus diesen Strukturen abgeleitet werden können. Die Autoren geben eine gründliche Einführung in den Gegenstand, die auch von Lesern ohne spezielle Kenntnisse über die Physik der Oberfläche verstanden werden kann. Sie vermitteln einen Überblick über den gegenwärtigen Stand des Gebietes auf fortgeschrittenem Niveau und unter Einbeziehung der neuesten Literatur (vor 1987). Das Buch wendet sich an alle, die an der modernen Physik von Festkörperoberflächen und -grenzflächen interessiert sind.

AKADEMIE-VERLAG BERLIN
Deutsche Demokratische Republik

FSC
www.fsc.org
MIX
Papier | Fördert gute Waldnutzung
FSC® C083411

Zeitfracht Medien GmbH
Ferdinand-Jühlke-Straße 7
99095 Erfurt, Deutschland
produktsicherheit@kolibri360.de